Chemical Feed Field Guide for Treatment Plant Operators

Calculations and Systems

by William C. Lauer, Michael G. Barsotti, and David K. Hardy

American Water Works Association

Chemical Feed Field Guide for Treatment Plant Operators:
Calculations and Systems
Copyright © 2009 American Water Works Association

AWWA Publications Manager: Gay Porter De Nileon
Senior Technical Editor: Melissa Valentine
Technical Editor/Project Manager: Martha Ripley Gray
Production Editor: Cheryl Armstrong
Cover Art: Daniel Feldman

Disclaimer
The authors, contributors, editors, and publisher do not assume responsibility for the validity of the content or any consequences of their use. In no event will AWWA be liable for direct, indirect, special, incidental, or consequential damages arising out of the use of information presented in this book. In particular, AWWA will not be responsible for any costs, including, but not limited to, those incurred as a result of lost revenue. In no event shall AWWA's liability exceed the amount paid for the purchase of this book.

Library of Congress Cataloging-in-Publication Data

Lauer, Bill.
   Chemical feed field guide for treatment plant operators : calculations and systems / William C. Lauer, Michael G. Barsotti, David K. Hardy.
      p. cm.
   Includes bibliographical references and index.
   ISBN-13: 978-1-58321-588-3
   ISBN-10: 1-58321-588-3
   1.   Drinking water--Purification--Materials. 2.   Drinking water--Purification--Mathematics. 3.   Feedstock. 4.   Materials handling--Mathematics. 5.   Engineering mathematics--Formulae. 6.   Chemical process control. I. Barsotti, Michael G. II. Hardy, David K. III. Title.

   TD451.L38 2008
   628.1'66--dc22

                              2008044305

**American Water Works
Association**

6666 West Quincy Avenue
Denver, CO 80235-3098
303.794.7711

Table of Contents

List of Figures

List of Tables

About the Authors

William C. Lauer

William C. Lauer is senior technical services engineer for the American Water Works Association (AWWA). Mr. Lauer has written and edited more than a dozen books and 50 articles and technical publications covering all aspects of the drinking water industry, including water quality, treatment, reuse, distribution system operation, management, and desalting. He is a recognized technical expert in the field and has consulted for NASA, US Environmental Protection Agency, the Government of Singapore, several major engineering design and construction firms, and many others in his more than 30 years in the drinking water supply field.

Michael G. Barsotti

Michael G. Barsotti is director of water quality and production for Champlain Water District in northwestern Vermont. Mr. Barsotti has written several papers addressing treatment facility optimization and has actively served the Partnership for Safe Water Program and the AWWA New England Water Works Filtration Committee. In addition, Mr. Barsotti has served several years as a state regulator. He has more than 22 years' experience in drinking water treatment plant operations and water quality management.

David K. Hardy

David K. Hardy is manager of the Utah Valley Water Treatment Plant for the Central Utah Water Conservancy District. David has been involved in the operations of water treatment plants since 1985. He served as a plant operator for 12 years and treatment plant manager for the past 11 years. He considers his most important title to be "operator" and continues to operate the plant, with frequency, to "feed his roots." His strengths are in his knowledge of process control and treatment plant optimization. The Utah Valley Plant and staff are recipients of the Partnership for Safe Water Phase IV Award of Excellence.

Foreword

This chemical feed operators' field guide is one of the "Field Guide" series of books published by the American Water Works Association (AWWA). These books are meant to be small, practical, how-to publications on specific subjects of importance to drinking water system operating personnel. These books omit most of the theory and background that have led to the use of the procedures described in the guides. This fundamental information is found in other AWWA publications and is referenced in the field guides.

This field guide, like the other books in this series, provides the information needed to do the work. Useful tables and easy-to-follow illustrations help system operators perform the chemical feed procedures described in the references. Field guides get straight to the point and provide the necessary information to perform the *most common* procedures. This approach leads the operator to the most useful solutions without needlessly complicating the issue with every possibility.

There are several special notes included throughout the book. Look for "Ops Tips" and the "Table Tamer" in callouts and text boxes for some help with important points and how to use some of the tables. There are also many "calculators" that help plug in the numbers needed to calculate a value.

"Calculators" provide easy-to-use, plug-in-the-numbers equations and examples that are used to give quick results. The conversion factors and other constants have been combined so that the equation is greatly simplified. The derivation of the formula is not given, just the result. More detail about the source of the constants shown for the calculators is shown in Appendix D.

Ops Tip

Operator Tips point out important points in the text.

Table Tamer   Illustrates how to use some of the more complicated tables.

Use calculator cX-x

| Calculation result and units of measure | = Constant × | Insert known value needed for the calculation |

Source water quality differences and treatment processes control the optimum chemical treatment at a given location. Accurately and precisely feeding water treatment chemicals is critical to production of high quality drinking water. The authors hope this information is helpful and is used to supplement hands-on training and thus becomes an indispensable companion for water treatment plant operations personnel.

William C. Lauer
Michael G. Barsotti
David K. Hardy

Acknowledgments

The authors wish to thank the following reviewers who provided the benefit of their experience to enhance this publication.

William Soucie, Central Lake County Joint Action Water Agency, Lake Bluff, Ill.

David Tuck, Greenwood Commissioners of Public Works, Greenwood, S.C.

Specific advice was provided regarding ozone feed calculations by Kerwin Rakness (Process Applications Incorporated). Suggestions and comments on chlorine dioxide feed were provided by Kevin Gertig and Grant Jones (both with the City of Fort Collins, Colo.).

Thank you also to the publication department at AWWA. The professionalism of our editors, Martha Ripley Gray and Melissa Valentine, greatly enhanced the quality of this field guide.

Chapter 1

Chemical Feed Quality Control

Water treatment plants use chemicals to remove contaminants, improve taste and appearance, and satisfy regulatory requirements. Accurate delivery of these chemicals to the point of application is critical to maintain and assess plant performance, to control the cost of treatment, and to predict the need for more supplies.

Many chemicals are used in water treatment. Table 1-1 lists the most common chemicals. The table groups chemicals by the most prevalent use, although some chemicals are used for several purposes.

Many chemical feed locations exist in a treatment plant. Rapid dispersion of coagulants occurs in the rapid mix chamber. Polymer filter aids are added just prior to filtration or even on top of the filters. Injecting disinfection chemicals after filtration reduces undesirable by-products. Stabilization of the water provides corrosion control as the water enters the distribution system (Figure 1-1). A careful match of the feed location with the correct chemical leads to peak treatment plant performance.

Quality Control Leads to Quality Assurance

Drinking water should be considered a food. As such, a water treatment plant is a food processing facility. Setting strict control measures at each step in the manufacturing process ensures the quality of the product. These steps usually include raw material specifications, control limits, personnel training and education, and product testing.

Ensuring the quality and accuracy of chemical application requires control procedures at each step, from selecting the correct chemical to ensuring peak dispersion of the chemical in the water. The main steps in chemical feed include

- Chemical selection and dosage determination
- Chemical purchasing specifications
- Quality screening prior to unloading from vendor
- Storage considerations to ensure chemical stability

Table 1-1   Common chemicals used in water treatment

Coagulation	Form	Fluoridation	Form
Aluminum sulfate (alum)	Liquid/dry	Calcium fluoride	Dry
		Sodium fluoride	Dry
Ferric chloride	Liquid/dry	Hydrofluosilicic acid	Liquid
Ferric sulfate	Dry	Sodium fluorosilicate	Dry
Ferrous chloride	Dry		
Polyaluminum chloride	Dry	**Softening**	**Form**
Sodium aluminate	Dry	Sodium chloride	Dry
Sodium silicate	Liquid/dry	Soda Ash	Dry
Cationic polymers	Liquid/dry	Calcium oxide (quicklime)	Dry
Anionic polymers	Liquid/dry	Calcium hydroxide (hydrated lime)	Dry/liquid
Disinfection Chemicals	**Form**	**Corrosion Control**	**Form**
Chlorine	Gas/liquid/generated*	Sodium hydroxide (caustic soda)	Liquid
Chlorine dioxide	Generated	Sodium polyphosphate	Dry
Ozone	Generated	Sodium hexametaphosphate	Dry
Chloramines	Generated		
Ammonia, anhydrous	Gas	Sodium tripolyphosphate	Dry
		Monosodium phosphate	Dry
Ammonium hydroxide	Liquid	Disodium phosphate	Dry
		Zinc orthophosphate	Dry
Ammonium sulfate	Liquid	Carbon dioxide	Gas
Sodium hypochlorite	Liquid/generated	Potassium hydroxide	Dry
		Calcium chloride	Dry
Hydrogen peroxide	Liquid	Phosphoric acid	Liquid
Sodium chlorite	Liquid		
Oxygen	Gas		
Dechlorination	**Form**	**Taste and Odor Control**	**Form**
Ascorbic acid	Dry	Powdered activated carbon	Dry
Calcium thiosulfate	Dry		
Sodium bisulfite	Dry	Granular activated carbon	Dry
Sulfur dioxide	Gas	Copper sulfate	Dry
Sodium metabisulfite	Dry	Potassium permanganate	Dry
Sodium sulfite	Dry		
Sodium thiosulfate	Dry		

*Generated: on-site generation or production of this chemical.

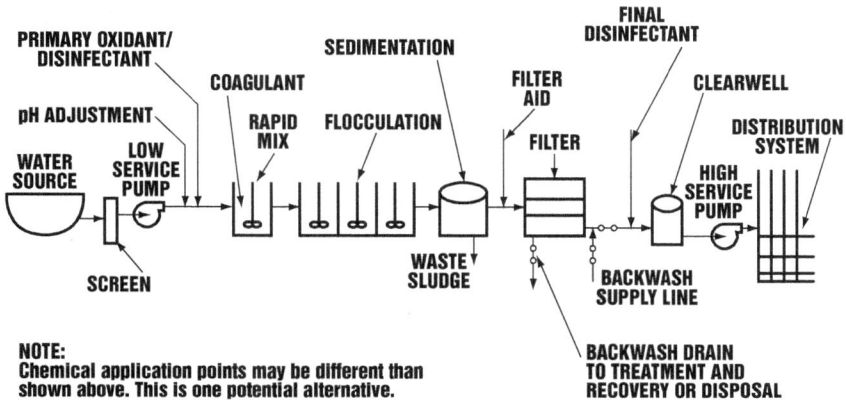

From Letterman, Raymond D., and American Water Works Association, Water Quality
and Treatment: A Handbook of Community Water Supplies. 5th ed. Copyright © 1999
McGraw-Hill. Reproduced with permission of the McGraw-Hill Companies.

Figure 1-1   Conventional water treatment plant with typical chemical
addition locations

- Equipment calibration
- Feed-rate adjustment and accuracy
- Chemical inventory control
- Chemical addition and mixing

Performing quality control procedures for chemicals helps assure
production of high-quality water from a water treatment plant. Controlling chemical handling and feed is first among the many steps of
plant operational control. The following sections discuss the quality
control procedures for each step.

Chemical Selection and Dosage Determination

Chemical selection depends on many factors. These factors include
the treatment objective, the quality and variability of the untreated
water, chemical availability, delivery time, suitable storage, feed
equipment and accuracy, safety considerations, cost, and employee
training. The treatment objective is perhaps the most important
factor in chemical selection. A treatment objective might include
reduction of lead levels in customers' homes with a corrosion inhibitor,
elimination of taste and odor compounds with ozone, or maintenance
of a chlorine residual at the farthest end of the system.

Untreated water variability may favor one coagulant over another.

Spring snowmelt may impact alkalinity in cold weather regions, and rain events may cause turbidity spikes in river supplies. Limited chemical availability may make some chemicals too expensive to consider. Delivery time may be critical to utilities with limited storage. Storage tank material and piping may not be compatible with some chemicals. For example, stainless steel fittings are not recommended for use with sodium hypochlorite.

Feed equipment may not be compatible with some chemicals. Chemical pump size may also dictate chemical selection. For example, corrosion inhibitors are often purchased in varying concentrations. Ideally, the selected chemical concentration will allow the pump to operate in its most efficient and accurate range. Safety considerations are often significant. For example, if proper storage for chlorine gas use is not available, sodium hypochlorite use might be considered. Cost is always a factor in chemical selection. Chemical selection may depend on the complexity of employee training or ease of use.

Dosage determination is based on treatment objective. For example, one objective for chlorine dose may be to achieve a target residual at the end of the distribution system. The objective for hydrofluosilicic acid addition may be to achieve a required concentration. The objective for coagulant dose may be to achieve some minimum turbidity goal.

Determining dosage targets is fundamentally based on achieving the treatment objective. Dosage targets can be based on chemical measurement results or on predictive test equipment estimates. Chlorine dose targets, for example, are set by monitoring the chlorine residual (chemical measurement) at a suitable sample site with on-line instrumentation or grab samples. The measured residual is compared to the applied chlorine dose. Over time, a relationship is established between the two values. The operator can then use this information to set an applied dosage that will achieve the desired residual.

The use of predictive test equipment, like a jar test apparatus, is useful for chemical dose determination and even chemical selection. Coagulation, for example, requires a specific chemical dose for best performance. Several different doses can be tested simultaneously in the jar test apparatus. Jar testing often determines the optimal dose in a shorter time than needed in the treatment plant. The use of predictive test equipment also avoids any problems that might occur when adjusting chemical dosages in the treatment plant.

Chemical Purchasing Specifications

Purchasing specifications for chemicals should contain requirements to ensure the correct chemical strength and grade is delivered. The specifications should specify the testing time needed to verify these requirements on delivery at treatment plant site. Special conditions added to bid or purchasing documents (see Appendix A: Chemical Specification Example Language) can ensure chemical reliability. Suppliers that sign these documents agree to the conditions allowing enforcement. Returning a delivery truck with a full load of rejected chemical sends a powerful message that stresses the commitment to quality.

The following are some suggestions of requirements for bid or purchasing documents:

- *Require meeting standards.* Use only chemicals certified to NSF Standard 60 or equivalent for water treatment chemicals. Chemicals should also meet AWWA standards. If there are no applicable AWWA or NSF standards, require the chemical to meet an American Society for Testing and Materials (ASTM) or another chemical industry standard. Some state regulatory agencies require certified chemicals. Check with the regulating authority for this information.
- *Receiving hours and duration.* Specify the hours and days of the week staff is available to receive shipments. On certain days or after normal work hours, personnel may not be available to unload the chemicals. Make sure to include enough unloading time (for example, three hours), so samples can be provided and any testing can be completed. Delivery drivers are typically eager to make their delivery.
- *Certificate of analysis.* Require load-specific (or perhaps a batch) certificate of analysis.
- *Shipment sampling responsibility.* Clearly identify who will be responsible for getting a sample of the shipment. (It is usually the truck driver, but it could be plant personnel.) Treatment plants usually provide a sample container. A clear (glass) container makes visual inspection easier. Some utilities require the supplier to provide a grab or "save" sample.
- *Wash-out documentation.* If chemicals are sent in a trailer or reusable container, specify that the trailer or container be dedicated for the use of water treatment chemicals. Alternatively, require a wash-out or clean-out certificate or

some other documentation that the contents of the previous
load were thoroughly removed.

- *Fill-line security.* Prohibit unloading until approved by plant
  personnel. Lock or secure all external fill-lines to prevent
  unloading before completing the quality control tests (Figure
  1-2). Clearly mark bulk chemical fill-lines with the chemical
  name.

- *Inspection prior to unloading.* Require inspection of chemical
  fill-lines and equipment used to offload bulk shipments.
  Make sure hoses are connected to the correct chemical
  fill-lines. Mixing some chemicals can result in dangerous
  conditions. Make sure that fill-lines, drums, and containers
  are not damaged or do not show evidence of leakage. Ensure
  the integrity of any tamper-proof security features.

  Require that all openings to the tanker be sealed with
  tamper-evident seals. The seals should be identified with
  numbers that are recorded on the bill of lading. Inspect the
  seals before unloading. Require the driver to don the proper
  safety equipment.

- *Testing prior to unloading.* Get a sample from every dry and
  liquid chemical delivery (compressed gases are exempt).
  Conduct the special tests included in the purchasing
  documents. These tests include, at least, appearance (for all
  liquid and dry chemicals) and specific gravity (for liquids).
  AWWA standards or other chemical industry standards list
  the appearance requirements, which may include color, odor,
  clarity, and visible foreign matter.

Specific gravity is tested using a hydrometer (Figure 1-3), an in-
expensive tool. A hydrometer consists of a glass tube with a weight
inside, at the bottom. Along the top stem is a graduated scale in
units of specific gravity. The hydrometer is gently immersed with
a slight spinning motion into a container containing the treatment
chemical. As the hydrometer buoys itself in the chemical and comes
to a rest, the depth to which it sinks is read off the scale as shown in
Figure 1-3. The value is the specific gravity. The temperature of the
chemical has a significant impact on the measured specific gravity.
The specific gravity requirement should be specified for the *delivery*

Ops Tip
Put all delivery tests in specifications so that they can
be enforced.

Courtesy of Bill Soucie, CLCJAWA

Figure 1-2   Locked chemical fill station

temperature. Warming or cooling samples to room temperature can take considerable time.

Specific gravity tables for most chemicals are available from the manufacturer (common liquid chemical specific gravity tables are shown in Appendix C). The test results should be recorded for every delivery for future reference. Rather than setting a strict limit on specific gravity, look for major changes from previous shipments.

Table 1-2 contains suggested limits for specific gravity screening tests of common water treatment chemicals. Treatment plants may need chemical strengths that differ from those shown in the table. Some plants may also need higher purity than that listed in Table 1-2. Each treatment plant or utility should set up its own screening tests and acceptance limits. These should be included in the chemical specifications to ensure enforcement.

Table 1-2   Examples of specific gravity requirements for common liquid chemicals

Delivery Screening Tests			
Chemical	Requirement	Specification	Specific Gravity Limit Example*
Liquid alum	% Al_2O_3 % Al	8.0% min 50% min	>1.31
Caustic soda	% NaOH	50% min	>1.52
Aqua ammonia	% NH_3	29% min	<0.90
Hydrofluosilicic acid	% H_2SiF_6	24% min	>1.20
Ferric chloride	% $FeCl_3$	40% min	>1.40
Sodium hypochlorite	% Cl_2	12% trade min	>1.16

*All specific gravity limits listed are at the delivery temperature. This may vary, so the limits are less precise than if the temperature of the test were controlled at 60°F.

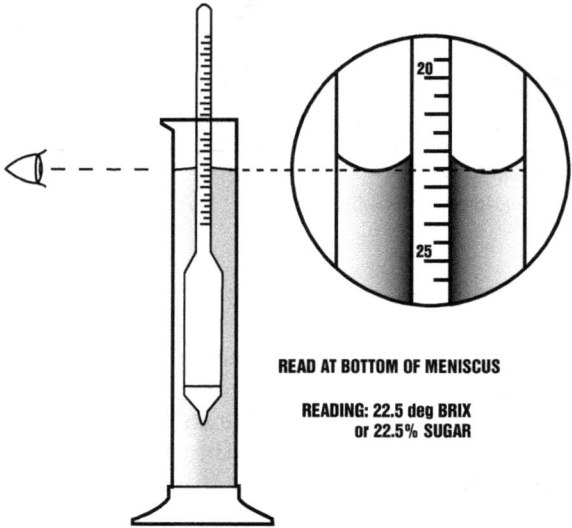

READ AT BOTTOM OF MENISCUS

READING: 22.5 deg BRIX
or 22.5% SUGAR

Figure 1-3   Reading a hydrometer

Storage Considerations Affecting Chemical Feed

Prolonged storage can adversely affect some chemicals. It is important to follow manufacturer (supplier) recommendations for proper storage. The strength and purity of some stored chemicals should be checked periodically to ensure that these values have not changed.

Table 1-3 lists the common chemicals used in water treatment and some of their storage issues. This is not a comprehensive list. The conditions listed to help stabilize these chemicals should be noted.

Feed Equipment Calibration or Verification

Calibration is the process of measuring the chemical delivered at various equipment settings. The measured output is then used to set the feeder to its needed output. This is necessary for accurate dosing when manually adjusting pumps that have speed or both speed and stroke dials. Calibration data can also be used to adjust digital settings such as maximum output.

Feed equipment verification involves comparing actual outputs to the expected outputs (spot-checking). Many treatment plants require chemical feeders (for example, metering pumps) to be verified on

Table 1-3   Common water treatment chemicals with storage issues

Chemical	Storage Issues	Suggested Storage Conditions
Chlorine gas	Leak containment	Containment building and gas scrubber
Sodium hypochlorite	Strength decay	Control temperature (less than 85°F) and dilute to reduce rate of decay, limit storage to less than 28 days, or monitor strength
Liquid alum	Strength decay	Control temperature; freezing point of 50% solution is about 2°F (−17°C)
Caustic soda	Solution strength decay and freezing	Dilute to 20–25% upon delivery, and reduce exposure to air
Aqua ammonia	Strength decay	Reduce evaporation by restricting vents
Hydrofluosilicic acid	Corrosive fumes damage equipment	Storage tanks should be ventilated directly outside
Polymers	Effectiveness decay	Control temperature

every shift. When the values consistently fail to meet the accuracy target (for example, <5 percent difference), calibration is warranted.

These practices ensure the accuracy of chemical feed equipment. All chemical feeders (even the most accurate, digitally controlled equipment) need regular, scheduled calibration.

Specific calibration and verification procedures for gas, liquid, and dry chemicals are discussed in detail later in this book. However, the following are some general considerations common for all chemical feed equipment (gas chemical feed equipment may be an exception):

1. Record the expected or indicated chemical feed rate.

2. Pump chemical from a graduated container into the treatment process or, alternatively, collect the pumped chemical while recording the time.

3. Record the amount of chemical delivered and the elapsed time to deliver to the treatment process.

4. Select another chemical feeder setting and repeat the process.

5. Graph the results. Divide the volume or weight by the time, and plot this value on the vertical axis and the feeder setting on the horizontal axis (Figure 1-4).

6. Digital feeders display the feed rate for each setting. These feeders must be checked to verify that they are feeding the display amount. Set the feeder and collect the chemical while recording the time. Convert this amount to the units on the feeder (for example, mL/min) and compare.

7. Pump outputs may also be verified by recording the weight delivered over a time period compared with the water pumped during the same period.

Feed Rate Adjustment and Equipment Accuracy

Calibrated chemical feed equipment ensures that a specific equipment setting will deliver the expected amount of chemical to the application point. Adjusting the chemical feed equipment is necessary to maintain a consistent dosage as treatment plant

Ops Tip

Liquid sodium hypochlorite can lose strength over time.

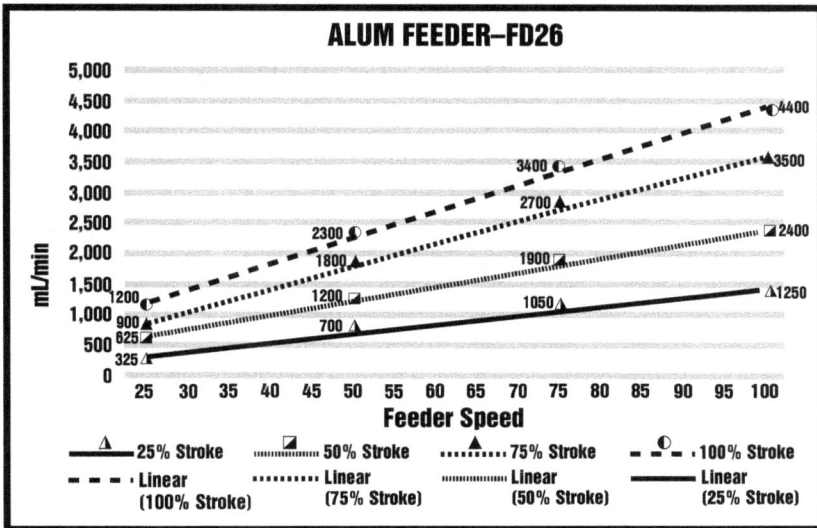

ALUM FEEDER–FD26

Courtesy of David K. Hardy, Central Utah WCD

Figure 1-4   Chemical feeder calibration chart

conditions change. The most common plant change is the production rate (plant output). Many treatment plants use control systems that automatically adjust the chemical feed equipment to match a change in plant production (flow pacing). However, some equipment needs manual adjustment.

Plants with system control and data acquisition (SCADA) systems may adjust chemical dosage automatically. The SCADA system communicates the feed setting to the feed equipment to achieve the needed chemical dosage. These systems also offer manual feed equipment control to achieve needed dosage changes. In either case, the dosage is determined first, and then the equipment is adjusted to deliver the correct amount of chemical to match the volume of water being treated (flow).

Alarms and other control measures may be used to assure correct chemical dosage. Flow-indicating switches are used to signal flowing chemical, sounding an alarm if flow stops. Chemical flowmeters are used to measure flow and calculate delivered dose. These meters will sound an alarm if the calculated dose falls outside a target range or changes suddenly. Some pumps have alarms that signal unexpected conditions.

Although the feed equipment is calibrated and set to deliver the correct amount of chemical, each system has its own accuracy limits. Every chemical feeder has a quantifiable error range. This is usually expressed as a percent (%) error. The equipment may deliver an amount above or below the set point (needed amount) equivalent to this percentage. Equipment delivering chemical with the needed accuracy should be selected.

The feed equipment is not the only source of error. The strength of the chemical may vary, thus introducing an error. When measuring the chemical strength, a percent error should be included, depending on the method used. Other parameters, such as temperature, altitude, pressure, and vacuum variability, may cause errors as well. All of these errors can combine in unpredictable ways to result in an amount delivered that is different from the intended amount.

Quality control procedures are used to lessen or at least understand the extent of errors introduced by chemical feed systems. Therefore, awareness of the error allows for its correction or compensation. The degree of accuracy of chemical feed can affect the cost of treatment when more chemical is used than is needed. Also, inaccurate feed systems can hide leaks and other losses. It is important to use equipment that has suitable accuracy.

Chemical Inventory Control

Keeping months of chemical inventory is often not possible, because of space limits, or desirable, because many chemicals degrade with time. However, running out of treatment chemicals is not a choice, and even running low is risky. Maintaining accurate chemical feed and records of historical chemical use is necessary for managing chemical inventory. To minimize costs, order *full-load* shipments.

Chemical Addition and Mixing

The efficient application of chemicals at the point of delivery has an impact on the effectiveness of the treatment. Inadequate mixing can cause inefficient chemical use, leading to the need to increase dosages. Control samples taken from locations where mixing is inadequate may provide misleading results. This leads to incorrect

Ops Tip

Good mixing is needed for efficient chemical addition.

treatment decisions. The chapters in this book for gas, liquid, and dry chemicals include specific suggestions for mixing and chemical application. Dispersing chemicals in the water evenly and quickly needs careful consideration. Efficient use of the chemicals depends this process.

Chemical Feed Locations and Timing

Water treatment plants use chemicals at many locations (and at other sites in a water system). Carefully locating the application point for a chemical can dramatically affect the chemical's effectiveness and treatment results. Figure 1-1 shows a conventional water treatment plant and some of the more common chemical application points.

Where multiple chemicals are used, the order and timing of chemical addition can be critical. For instance, polymer coagulation aids are sometimes used with alum. Many operators report the order of addition (polymer before or after alum addition) has an effect on performance. Some operators even express the importance of the time between adding the two chemicals. These chemical addition details need to be explored at each site. Jar testing is sometimes effective when comparing these variations. Only promising treatment sequences (from jar testing) are then evaluated in the plant full-scale.

Quality Control at Each Step

An entire system of checks comprises the quality control of chemical feed in water treatment. Employing this system reduces the risk of contamination, ensures efficient treatment, lowers the cost of treatment, and ensures the quality of the drinking water produced.

Chapter 2

Chemical Feed—
Common System Components

All chemical feed systems (gas, liquid, dry) have several components in common (Figure 2-1). All chemicals must be conveyed (delivered) to the process water flow, dispersed into the water, and their dosage controlled consistently and accurately.

Conveyance

Conveying the chemical to the point of application is a critical part of the chemical feed system. Usually, it is best to position the chemical feed equipment as close to the point of application as possible. This reduces problems that might occur in transport and simplifies feed control. However, many times chemicals must be stored a distance from the application point. Conveyance then becomes an important consideration.

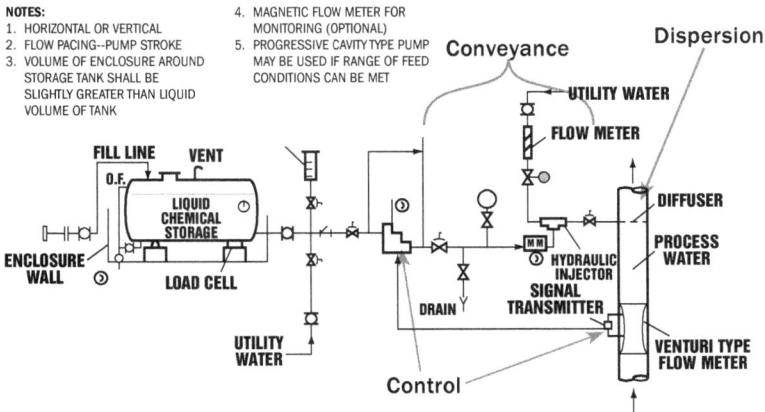

Adapted from Fig. 4.1-2 in Kawamura, Susumu, Integrated Design and Operation of Water Treatment Facilities, *2nd ed. Copyright © 2000 John Wiley & Sons, Inc. Reprinted with permission of John Wiley & Sons, Inc.*

Figure 2-1   Chemical feed system, common components

The chemical is moved from the feeder to the application point in several ways. It is rare for a dry chemical to be fed directly into the process water. Where this is done, aggressive mixing is applied at the point of application. It is much more common for the dry chemical to be delivered as a solution. This is accomplished by gravity feed of the solution, a chemical pump, or use of an injector (*eductor* or *ejector*).

Gravity solution feed systems usually drop the chemical into the process stream from a single pipe or a perforated pipe distributor. The chemical solution is normally prepared from dry chemical or diluted liquid chemical (Figure 2-2). Gases are not often conveyed to the point of application by gravity. Where this is practiced, the gas is first dissolved in water to create a solution. This method is not common because dosage control of the solution is difficult.

Once the chemical is measured (metered) to prepare the proper strength solution, a motive pump can be used to move it to the point of application. Motive pumps may be metering pumps but not always. Simple centrifugal pumps (or other types) can be used to provide satisfactory pressure and flow to deliver the solution to the application point. It is necessary to use care when selecting these pumps to ensure the materials are compatible with the diluted chemical being transported.

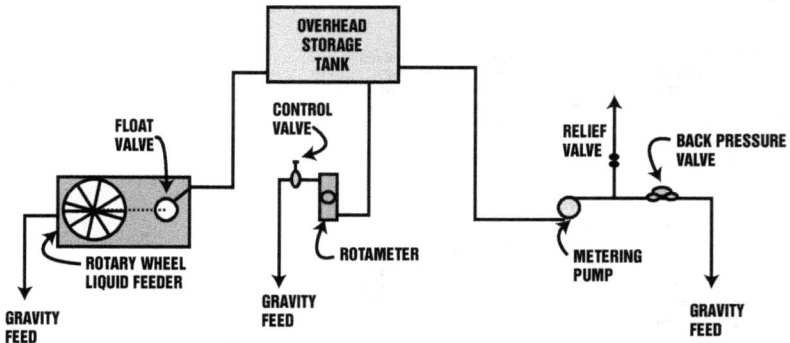

Figure 2-2   Liquid chemical gravity feed system

Ops Tip

The terms *injector*, *ejector*, and *eductor* are often used interchangeably.

Ops Tip
Liquid alum can hydrolyze at pH above 3.5. Keep the motive water ratio at about 5:1 to prevent this.

Motive water (or dilution water) is sometimes used to move the metered chemical to the point of application (Figure 2-3). Water pressure is often supplied from the treatment plant potable water system with proper backflow prevention (sometimes untreated water is also used). Advantages of motive water systems include shortened transport time and improved mixing. Disadvantages include possible changes in the chemical properties (caused by dilution) and added demand on the treatment plant service water system.

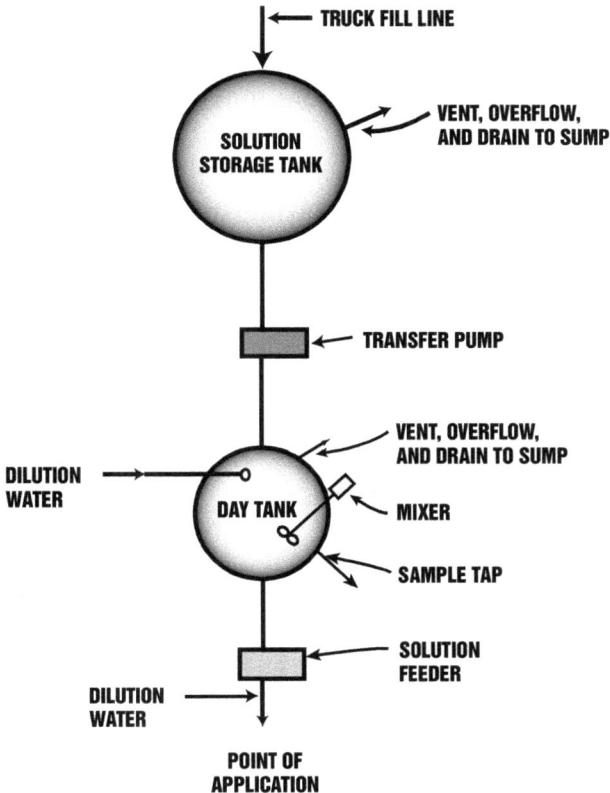

Figure 2-3   Liquid chemical feed system using dilution water

When applying chemicals, it is important to establish safeguards to prevent siphoning of too much chemical during unusual vacuum conditions. Examples of siphoning controls are three-way valves, backpressure valves, or nonflooded suction-type installations.

Venturi-type ejectors (Figure 2-4) are popular for the delivery of chemical solutions. These devices use water pressure through a Venturi to produce a vacuum that draws the chemical solution into the water stream. This stream is then directed into the process water to deliver the chemical. Ejectors are simple and reliable as long as a pressurized water supply is available. It is important to use materials for all of these devices that are compatible with the chemicals being used.

Chemical gases are often delivered to the application site using a Venturi injector (eductor, Figure 2-4). The gas is drawn from the storage vessel using the vacuum created as water flows through the device. The gas mixes with the water inside the injector throat. The injector water is then introduced into the water being treated (either inside a pipeline or using a diffuser).

Dispersion

Efficiently blending the chemical into the process water is often an overlooked but crucial process that can affect the effectiveness and cost of chemical treatment. Many chemical applications need almost instantaneous dispersion for maximum effectiveness. Depending on the chemical properties (that is, viscosity, miscibility, concentration), this can require high-intensity mixing or gentle blending with the process water.

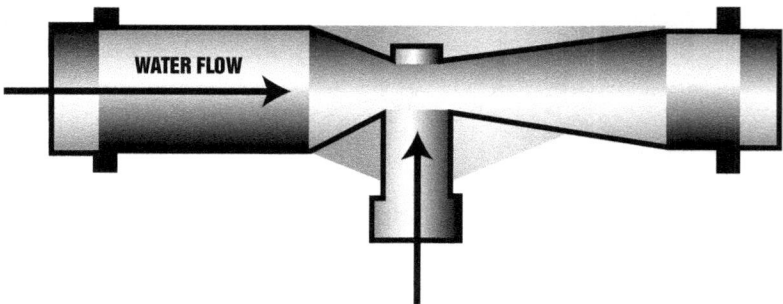

Figure 2-4   Venturi injector

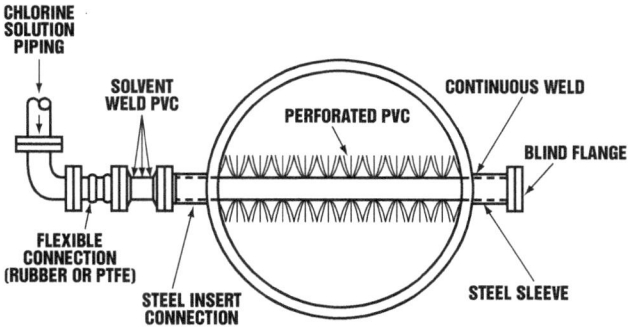

Figure 2-5   Perforated pipe diffuser

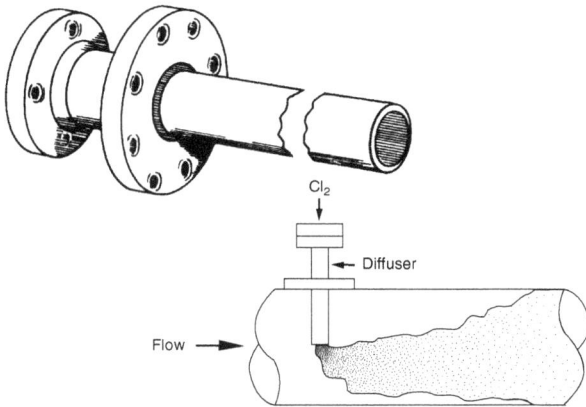

Figure 2-6   Simple pipe diffuser made from Schedule 80 PVC

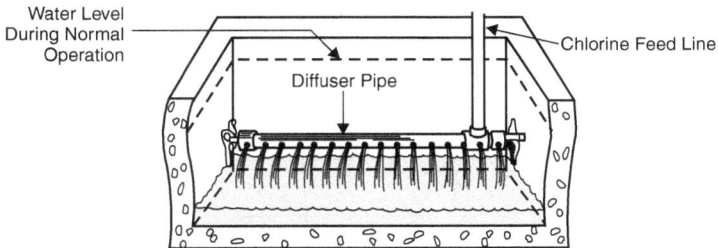

Figure 2-7   Open channel diffuser

Diffusers are used to spread the chemical and to provide mixing energy. There are many different types of diffusers. Ozone and carbon dioxide feed systems commonly use fine bubble diffusers. Chlorine diffusers (Figures 2-5, 2-6, and 2-7) are often perforated pipe or even a simple open pipe. Many liquid chemicals are delivered into the process flow with diffusers. It is important to spread the chemical evenly in the water and to provide adequate mixing.

These pipe diffusers should be carefully made so all the chemical solution is not fed preferentially through the holes nearest to the source. Depending on the volume being delivered to the diffuser and the pressure, this may not be an issue. Occasionally, however, graduated size and hole spacing may be needed.

Mixing

It is often important to mix the chemical quickly into the process stream to get the best results. Many types of mixing and dispersing systems are designed to achieve the needed result. Several mixing systems are pictured in Figures 2-8 through 2-10. Pumps are sometimes used as mixers either by circulating the chemical solution into the process stream or by providing turbulence at the point of application.

Mechanical mixers are not used in some applications because of site limitations or because other methods are satisfactory for the specific chemical. Examples of nonmechanical chemical application systems are diffusers of various kinds, as illustrated in Figure 2-11.

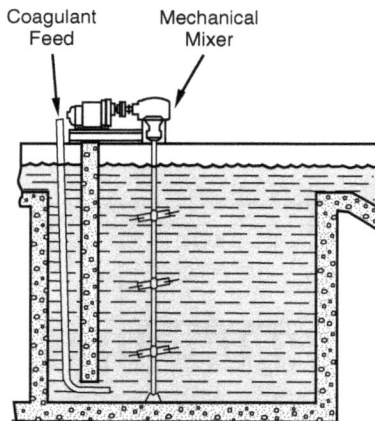

Figure 2-8   Multiple-blade mechanical mixer

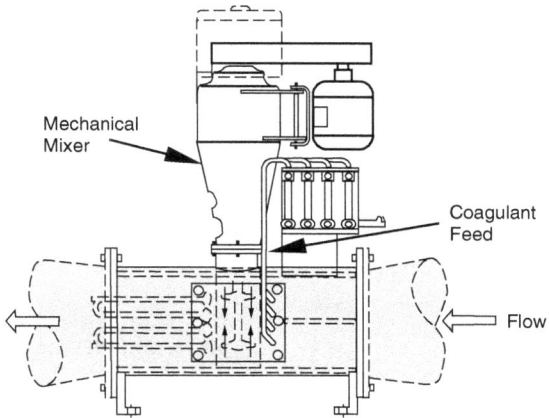

Figure 2-9   In-line mechanical mixer

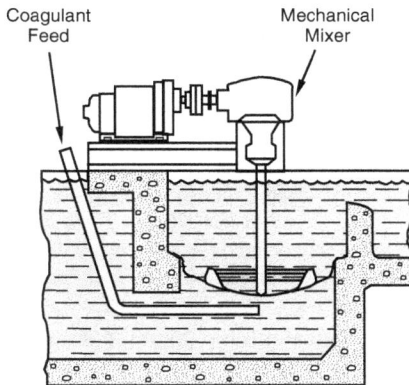

Figure 2-10   Single-blade mechanical mixer

(A) PORTION OF INFLUENT
FLOW USED TO DISPERSE
CHEMICAL INTO BULK FLOW

CHEMICAL

RAPID
MIXING

PUMP

WATER TO BE MIXED
WITH CHEMICAL

MIXED WATER
AND CHEMICAL

DIFFUSER PLATE

(B)   CHEMICAL

MIXED WATER
AND CHEMICAL

WATER TO BE MIXED
WITH CHEMICAL

IN-LINE VANES RESULT
IN TURBULENCE AND
RAPID MIXING

(C) CONVERGING SECTION RESULTS   CHEMICAL
IN INCREASED FLUID VELOCITY

DIVERGING SECTION
ALLOWS FOR PRESSURE
RECOVERY AND MIXING

WATER TO BE MIXED
WITH CHEMICAL

MIXED WATER
AND CHEMICAL

LOW-PRESSURE IN
THROAT SECTION DRAWS
IN CHEMICAL

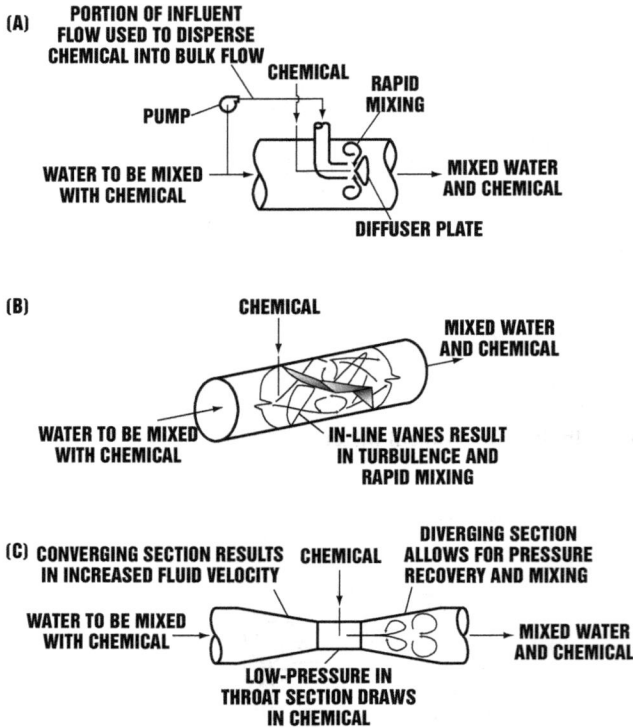

From Crittenden, John, Montgomery Watson Harza, et al., Water Treatment Principles and Design. *2nd ed. Copyright © 2005 John Wiley & Sons, Inc. Reprinted with permission of John Wiley & Sons, Inc.*

Figure 2-11   Types of chemical mixers: (a) pump diffusion, (b) static, (c) venturi

Dosage Control

Dosage control systems are used for liquid, dry, and gaseous chemicals. Many chemical feed controls are managed by computerized SCADA systems (Figures 2-12 and 2-13). These systems can be integrated with laboratory information management systems and utility data management and control systems. Manual or semiautomatic control systems are common and can provide accurate chemical feed. Record keeping may require considerable labor and can sometimes introduce errors.

Photo by Bryan Bechtold, AWWA

Figure 2-12   Computer SCADA control screens

Photo by Bryan Bechtold, AWWA

Figure 2-13   Large control center for public utilities

Chemical feed control strategies include manual, on–off constant rate, feedforward, feedback, and combinations of these methods. Selecting the best control strategy depends on several factors, such as the degree of expected variation in flow or water quality, the availability of automated equipment and computer systems, the chemicals in use, and the cost of the system. Each strategy is discussed in more detail in the following sections.

Manual Control

Chemicals are added continuously and at a constant rate. Adjustments are often made by plant operators at fixed intervals, such as once a shift or once a day. These adjustments include pump stroke length or frequency, strength of chemical solution, and valve position.

Manual control is most suitable for applications in which chemical control is not critical. Manual control also may be acceptable when established control ranges are wide or where treatment conditions do not change rapidly.

On–Off Constant Rate

A chemical feed pump (or other constant-rate feed device) is automatically cycled on and off by a control signal. This method is applicable to systems that do not need continuous chemical feed (for example, periodic oxidant feed to water intake to reduce zebra mussels). This control method is not common in water treatment.

Closed Loop

These systems are designed to detect changes in chemical demand and to compensate for them to control the system. Typical uses are to adjust corrosion inhibitor feed rate (based on changes in water temperature) or coagulant feed rate (based on influent turbidity readings).

Ratio control or flow pacing is a form of feedforward control in which the chemical pump or other metering device is automatically adjusted in proportion to a variable, such as water flow rate (flow pacing, Figure 2-14). Ratio control is most often used to maintain a fixed concentration of chemical in a water stream where the flow rate varies.

The primary disadvantage of this control scheme is lack of on-line feed confirmation. Although the controller sends a signal to the pump, there is no guarantee that the metering pump output matches

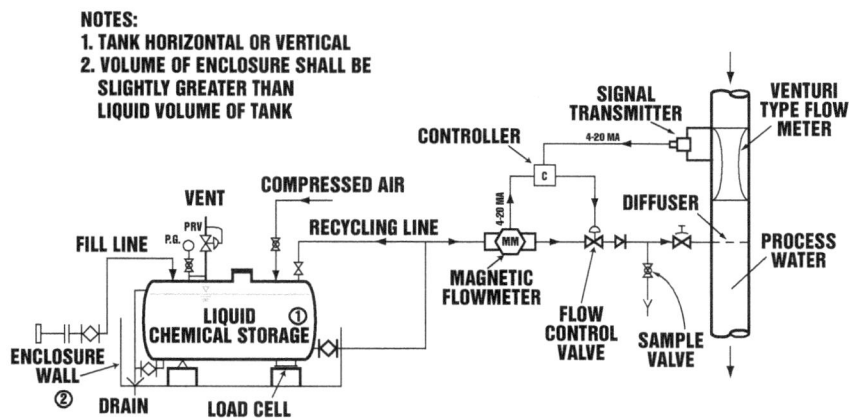

NOTES:
1. TANK HORIZONTAL OR VERTICAL
2. VOLUME OF ENCLOSURE SHALL BE SLIGHTLY GREATER THAN LIQUID VOLUME OF TANK

From Kawamura, Susumu, Integrated Design and Operation of Water Treatment Facilities, 2nd ed. Copyright ©2000 John Wiley & Sons, Inc. Reprinted with permission of John Wiley & Sons, Inc.

Figure 2-14   Flow pacing (feedforward) control system

the controller signal or that the metering pump is working. This disadvantage is lessened by pump output signals and alarms. Instrumentation is also available to provide this information so that verification is possible when using this extra equipment. For example, chemical flow switches and flowmeters can be used to ensure correct dose.

Another closed-loop control system monitors the value of the controlled variable and continuously compares this value with the desired value. When the detected value varies from a predetermined set point, the controller produces a signal indicating the degree of deviation. In many water treatment applications, this signal is sent (Figure 2-15) to a metering pump and the pump's output is automatically changed.

One of the most common examples of this control system is the feed of chlorine (liquid sodium hypochlorite) where a chlorine analyzer measures the residual concentration. When the controller detects a difference between the set point and the measured chlorine residual, it changes the pump speed or valve position to adjust chlorine residual to the set point.

The main disadvantage of this control system is that control action does not occur until an error develops. This potential problem can be mitigated with computer logic controllers. Another key problem is the dependency on the analyzer signal. In many systems, analyzer accuracy and reliability are questionable. Attentive analyzer maintenance and calibration can relieve this issue.

An automated closed-loop system includes both flow pace and analyzer value set-point adjustments. Water treatment chemical feed control strategies commonly employ a combination of these methods. The chemical feeder, for example, is set to deliver the required dosage. As the process flow changes, the feeder is adjusted automatically to match the flow. A sensor may also be in place to measure a particular character of the chemical. Sensing a variance from the set point, a signal can also be sent to the feeders to maintain the set amount. The two signals are managed within the computer control system to maintain a consistent dosage.

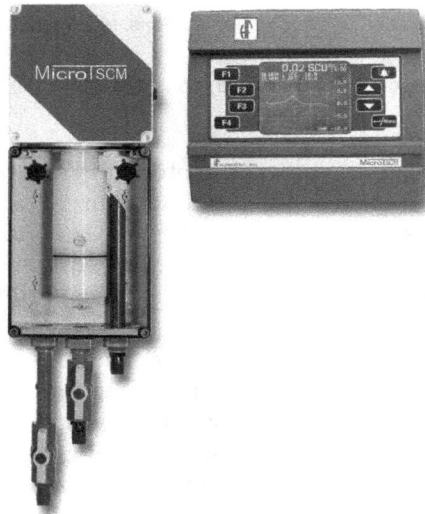

Manufactured by HF scientific

Figure 2-15   MicroTSCM streaming current analyzer

Chapter 3

Chemical Feed of Gases—
Equipment and Systems

Only a few gases are commonly used in water treatment. These include chlorine, oxygen, carbon dioxide, sulfur dioxide, anhydrous ammonia, chlorine dioxide, and ozone. Descriptions of gas feed systems and equipment are, therefore, limited to these specific chemicals. Other gases, although not specifically discussed in this field guide, may use similar feed systems. This field guide is not a design manual; for that, see *Water Treatment Plant Design* (Baruth 2005, chapter 9). It therefore does not include some system details. Site-specific conditions may also dictate differences in feed systems. However, this chapter does show typical system layouts used in many locations.

Chemical Details

Each chemical has specific characteristics that affect storage and feed system design. Table 3-1 lists some information about the gases used in water treatment. A brief summary of the uses for each chemical follows.

Chlorine

Chlorine is the most common disinfectant used in water treatment. This versatile chemical is not only used as a primary and secondary disinfectant but also as an oxidant of organic and inorganic substances. In addition, chlorine used before filtration can significantly improve particulate removal.

Chlorine, when combined with ammonia, forms chloramines. This form of chlorine is a common secondary disinfectant used in

Ops Tip Primary disinfection is used to satisfy requirements for inactivation of bacteria and viruses. Secondary disinfection is also known as *residual disinfection*. A residual is maintained within the distribution system to ensure the safety of water delivered to customers.

Table 3-1   Common gases used in water treatment

Chemical, Formula	Common Name	Shipping Containers	Solubility in Water, lb/gal	Commercial Strength, %
Chlorine, Cl_2	Chlorine gas, liquid chlorine	100 lb, 150 lb, 1-ton cylinders; 15-, 30-, 55-ton tank cars	0.07 (60°F)	99.8 (Cl_2)
Carbon dioxide, CO_2	Liquid carbon dioxide	Bulk tank cars	0.012 (77°F)	99.5
Oxygen, O_2	LOX	Cylinders, tank trucks, rail cars	3.16 (77°F)	99.5
Anhydrous ammonia, NH_3	Ammonia	100 lb, 150 lb, 1-ton cylinders; 15-, 30-, 55-ton tank cars	3.1 (60°F)	99.9 (NH_3)
Sulfur dioxide, SO_2	Sulfur dioxide	100 lb, 150 lb, 1-ton cylinders	11.3 (68°F)	100 (SO_2)
Ozone, O_3	Ozone gas	Generated on-site by electric discharge through dry air or oxygen	8.3 mg/L (68°F)	1–5 (produced on-site)
Chlorine dioxide, ClO_2	Chlorine dioxide gas	Generated on-site from chlorine gas and sodium chlorite	70 g/L (68°F)	75–100 (produced on-site)

water supply distribution systems. Chlorine can also serve as one of the chemicals needed to produce chlorine dioxide (another disinfectant or oxidant).

Even though chlorine is toxic in the concentrated gaseous form, safe handling procedures and its use as a disinfectant for drinking water are well understood. It is necessary to employ site-specific strategies to reduce forming undesirable by-products in the treated water.

Carbon Dioxide

Carbon dioxide is used for pH adjustment. When added to water, it forms carbonic acid (a mild acid). It neutralizes high-pH water resulting from lime-soda softening (*recarbonation*) and adjusts alkalinity of naturally high-pH water.

Oxygen

Oxygen is a mild oxidant used in the production of ozone. Ozone is a powerful oxidant and primary disinfectant. Ozone is created when electrical discharge passes through filtered and dried air or oxygen gas (liquefied oxygen [LOX]). Using atmospheric air as the oxygen supply produces a lower concentration of ozone than when using liquid oxygen. This result may justify the added cost of this chemical.

Anhydrous Ammonia

The primary use of anhydrous ammonia (ammonia gas) in water treatment is to combine with chlorine to form chloramines. Chloramines are used both as primary and secondary disinfectants. Use as a secondary disinfectant (residual in the distribution system) is more common. A typical treatment strategy is to use free chlorine to satisfy the USEPA regulatory CT requirements[*] as a primary disinfectant. Ammonia is then added to combine with the free chlorine residual to form chloramines for use as the secondary distribution system disinfectant. The ammonia added is carefully controlled to ensure that all the free chlorine is combined and little free ammonia remains. This control is necessary because the presence of free chlorine can form regulated by-products. Free ammonia can increase the growth of nitrifying bacteria, thus causing residual demand that could lead to conditions that could violate the Total Coliform Rule.

Sulfur Dioxide

It is not common to use sulfur dioxide in drinking water treatment. However, it is used to dechlorinate highly chlorinated water or to quench the chlorine residual before a chlorine sensitive treatment process. Because of the possibility of forming by-products, practicing breakpoint chlorination (high chlorine dosages to destroy ammonia; see *Water Treatment Operator Handbook* [Pizzi 2005]) is not as common as in the past. There are situations, though, where this approach is acceptable. Although unusual, quenching of the residual chlorine can be carried out with sulfur dioxide (for example, before membranes that are sensitive to chlorine).

---

[*]   CT is defined as the product of the residual disinfectant concentration, C, in mg/L, and the contact time, T, in minutes, that the residual disinfectant is in contact with the water.

Ozone

Ozone is a powerful oxidant and primary disinfectant. It is unstable, so storage is impossible. It is, therefore, produced as needed and on-site. Ozone is produced by passing an electrical discharge through air or oxygen. This is one of the major uses for oxygen in a water treatment plant.

As an oxidant, ozone removes iron, odors, and sometimes color, from water. It is used to oxidize organic substances in water, so they can be more easily removed in subsequent processes, i.e., coagulation, sedimentation, and filtration. Occasionally, ozone is combined with peroxide to increase its oxidation potential (peroxone). Peroxone can remove a wide variety of organic contaminants.

Ozone can be an effective method of inactivating *Cryptosporidium*. This has led to an increasing interest in the use of ozone for disinfection. It can be effective against both bacteria and viruses as well. Ozone does not form chlorinated by-products, but it can produce bromate (a regulated substance) if bromide is present. Another issue with ozone is that it rapidly decays, leaving no protective residual. In addition, ozone may produce biologically available organic carbon, potentially fueling distribution system bacterial regrowth. Therefore, ozone is commonly used with biologically active filtration to reduce this drawback. Water disinfected with ozone requires the application of a secondary disinfectant (such as chlorine or chloramines) to satisfy distribution disinfection requirements.

Chlorine Dioxide

Chlorine dioxide is prepared on-site as needed because the compressed liquid form is explosive at normal room temperature. There are several alternative production methods (using the reaction between chlorine gas and sodium chlorite). Chlorine dioxide is commonly delivered to the application point as a solution.

Chlorine dioxide is a selective oxidant and a powerful primary disinfectant. It does not combine with ammonia, like chlorine, so it is useful as a disinfectant when ammonia is present. The use of chlorine dioxide instead of chlorine results in fewer chlorinated by-products. And it does not form bromate like ozone. It does, however, produce chlorite as a reaction product, and this substance is regulated (maximum contaminant level [MCL] 1.0 mg/L). Therefore, the dosage of chlorine dioxide is controlled (maximum residual 0.8 mg/L), thus limiting producing chlorite from this source.

Oxidation of manganese is one of the main uses of chlorine dioxide. Another advantage of chlorine dioxide is its effectiveness in inactivating *Cryptosporidium*. Although it is sometimes used as a residual disinfectant in the distribution system, it is more common to use chlorine dioxide only as a primary disinfectant and then use chlorine or chloramines to provide a distribution system residual.

Delivery and Handling

Gases are contained in pressurized vessels such as rail tank cars, tank trucks, or cylinders. The chemicals are delivered to the user either in these vessels or transferred from bulk storage to cylinders. Rail tank cars and bulk tank trucks can connect directly to the treatment plant feed system. Cylinders are removed from delivery trucks and stored on-site until needed. Some cylinders are, therefore, connected to the feed system and some are in storage (Figure 3-1).

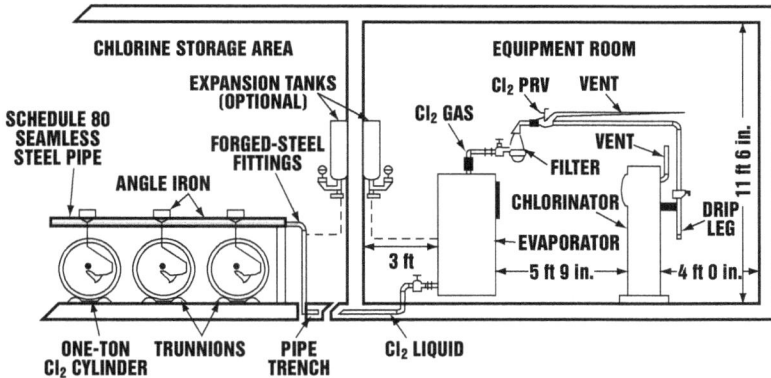

From Handbook of Chlorination and Alternative Disinfectants, *4th ed. by Geo Clifford White, copyright © 1998. Reprinted by permission of John Wiley & Sons, Inc.*

Figure 3-1   Chlorine cylinder connection illustration

Chemical Feeders

Chemical feeders for gases are remarkably similar regardless of the chemical. A chlorinator can feed sulfur dioxide, anhydrous ammonia, and carbon dioxide with only minor adjustments. Carbon dioxide, for example, is contained at a higher pressure. This condition requires installing a pressure-reducing valve between the supply and the vacuum regulating valve. For ammonia feed, the injector (eductor) water hardness must be below 35 mg/L. Chlorinators can feed oxygen, but this is not common because LOX storage and feed systems are usually self-contained. The main parts of a gas feeder (Figure 3-2) include vacuum regulator, pressure check and relief valve, rotameter and rate adjustment, differential regulating valve, relief valve, vacuum gauge, and injector.

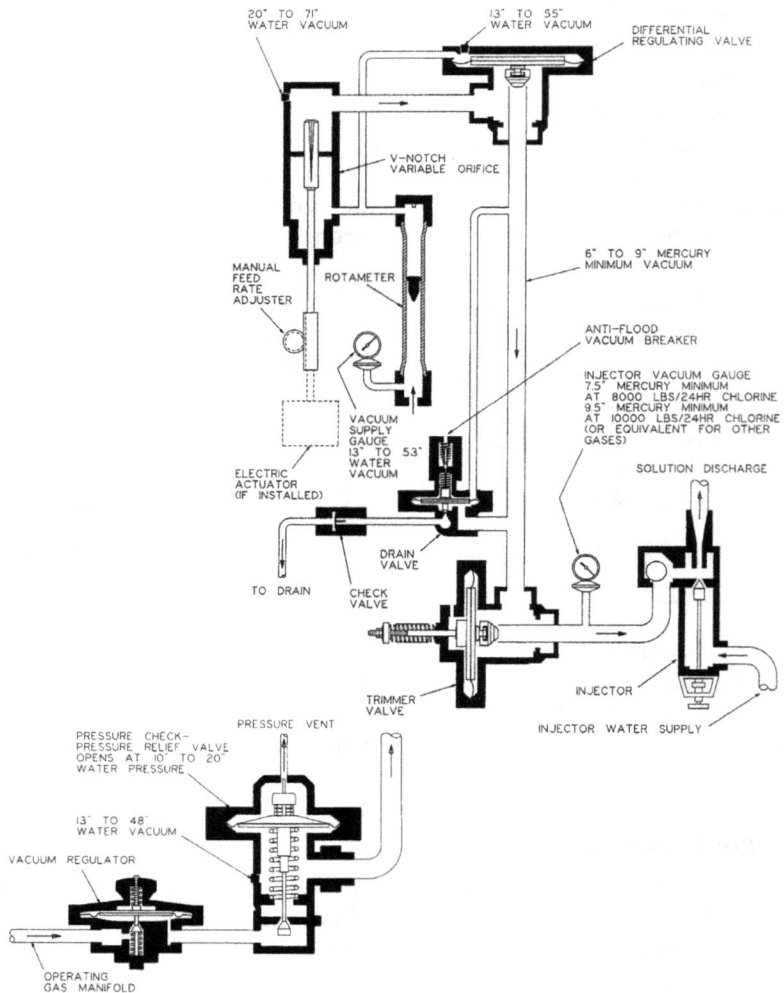

Courtesy of Siemens Water Technologies

Figure 3-2   Cabinet chlorinator

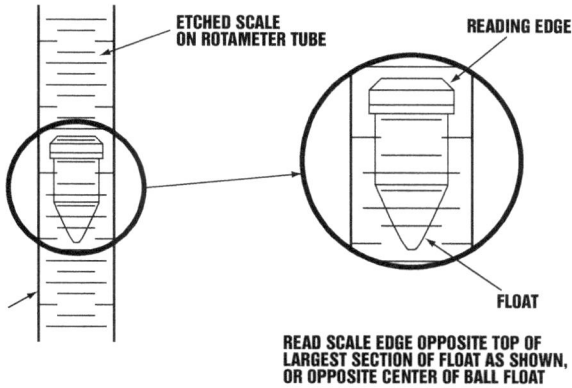

Figure 3-3   Rotameter close-up

Gas Feed Systems

All the units linking chemical delivery, storage, feed, and control comprise the feed *system*. There are many possibilities to consider when selecting from all the choices to design the best system. The gas feed system schematic gallery (Figures 3-4 through 3-8) shows some of the most common arrangements.

Gas Feed System Gallery

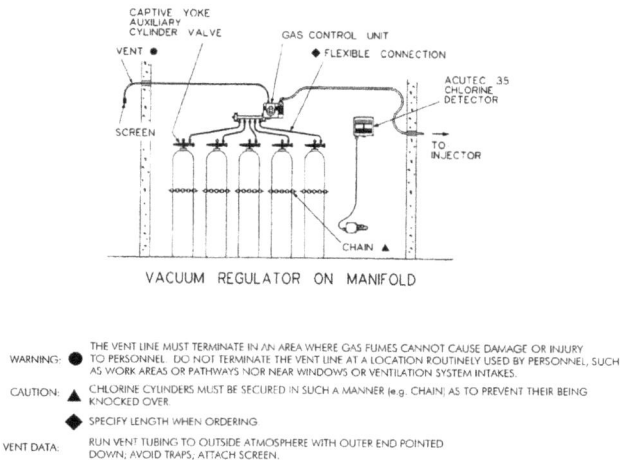

Courtesy of Siemens Water Technologies

Figure 3-4   150-lb cylinder chlorination system

VACUUM SEAL
O-RING

LEAD
GASKET

TO VENT

CHLORINE
CYLINDER VALVE
STEM

INLET
SAFETY
VALVE

RATE
VALVE

OUTLET
CONNECTION

CHLORINE
CYLINDER VALVE
PACKING UNIT

VENT VALVE

VACUUM
LINE

YOKE
CLAMP

RATE INDICATOR

INLET FILTER

REGULATING
DIAPHRAGM
ASSEMBLY

EJECTOR AND
CHECK VALVE
ASSEMBLY

CHLORINE
GAS

WATER SUPPLY
TO EJECTOR

EJECTOR
DISHARGE

CHLORINE
LIQUID

CHLORINE
CYLINDER

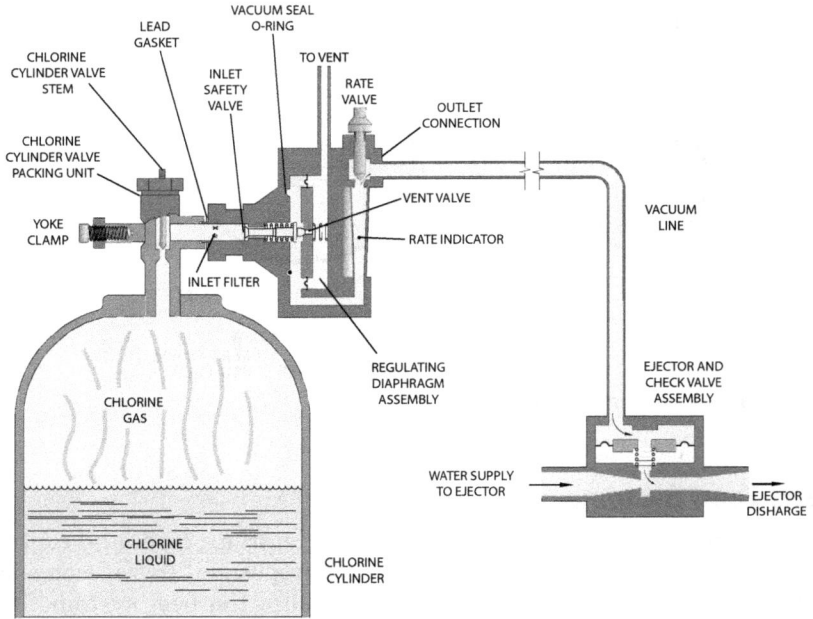

Courtesy Severn Trent Services

Figure 3-5   Cylinder-mounted chlorinator

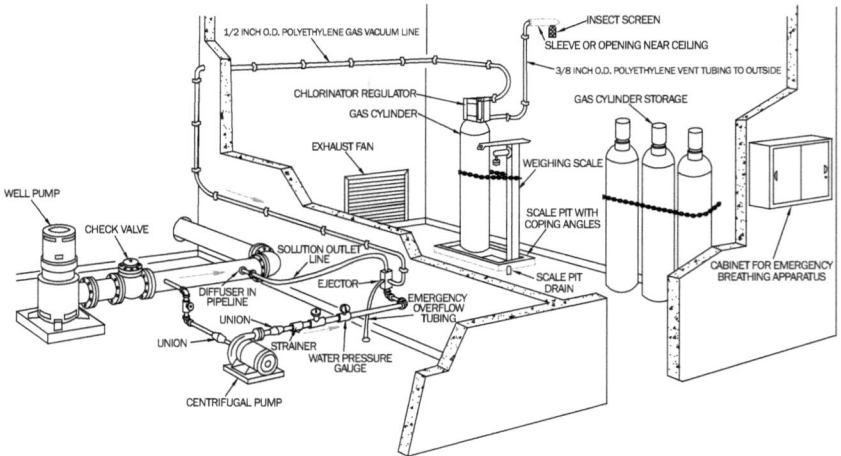

1/2 INCH O.D. POLYETHYLENE GAS VACUUM LINE

INSECT SCREEN

SLEEVE OR OPENING NEAR CEILING

3/8 INCH O.D. POLYETHYLENE VENT TUBING TO OUTSIDE

CHLORINATOR REGULATOR

GAS CYLINDER

GAS CYLINDER STORAGE

EXHAUST FAN

WEIGHING SCALE

WELL PUMP

SCALE PIT WITH
COPING ANGLES

CHECK VALVE

SOLUTION OUTLET
LINE

CABINET FOR EMERGENCY
BREATHING APPARATUS

DIFFUSER IN
PIPELINE

EJECTOR

SCALE PIT
DRAIN

EMERGENCY
OVERFLOW
TUBING

UNION

UNION

STRAINER

WATER PRESSURE
GAUGE

CENTRIFUGAL PUMP

Courtesy Severn Trent Services

Figure 3-6   Well chlorination system

REMOTE FLOWMETER WITH RATE VALVE   VACUUM TUBING

AUTOMATIC SWITCHOVER MODULE   VACUUM TUBING

EJECTOR

RUBBER HOSE

WATER INLET

VACUUM REGULATOR NO. 1   VACUUM REGULATOR NO. 2

VENT (TYP)

GAS CYLINDER NO. 1   GAS CYLINDER NO. 2

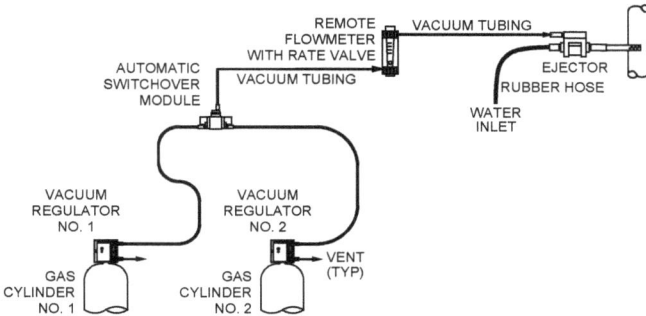

Courtesy Severn Trent Services

Figure 3-7   Cylinder-mounted automatic switchover system

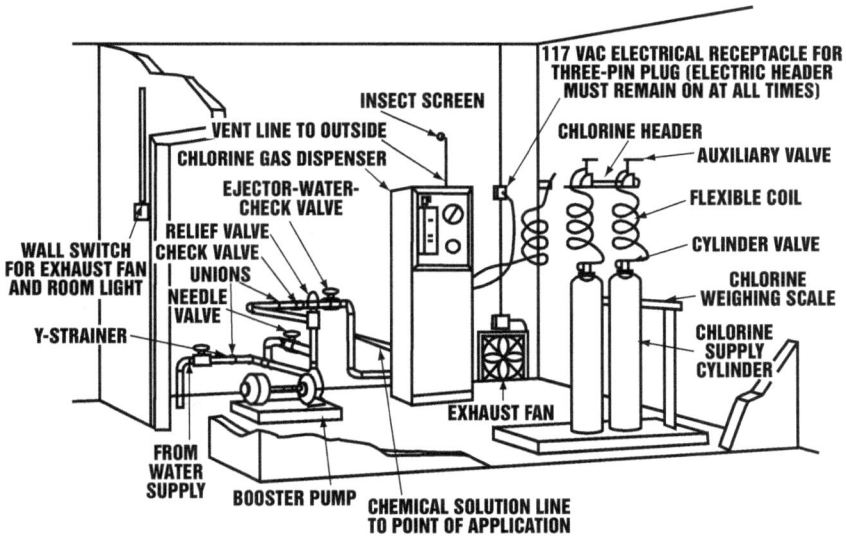

117 VAC ELECTRICAL RECEPTACLE FOR THREE-PIN PLUG (ELECTRIC HEADER MUST REMAIN ON AT ALL TIMES)

INSECT SCREEN

VENT LINE TO OUTSIDE

CHLORINE GAS DISPENSER

CHLORINE HEADER

AUXILIARY VALVE

EJECTOR-WATER-CHECK VALVE

FLEXIBLE COIL

RELIEF VALVE

WALL SWITCH FOR EXHAUST FAN AND ROOM LIGHT

CHECK VALVE

UNIONS

NEEDLE VALVE

CYLINDER VALVE

CHLORINE WEIGHING SCALE

Y-STRAINER

CHLORINE SUPPLY CYLINDER

EXHAUST FAN

FROM WATER SUPPLY

BOOSTER PUMP   CHEMICAL SOLUTION LINE TO POINT OF APPLICATION

Figure 3-8   Chlorination storage and feed room

Ozone Systems

On-site ozone generators operate by passing an electrical discharge through dry air or oxygen. Ozone produced is then bubbled through the water in a specially designed contactor or educted into a side stream of water that is then mixed with the water to be treated. Any residual ozone escaping from the water into the air is trapped and converted to oxygen before it is vented. Safe use of ozone equipment requires ozone air monitors to ensure ozone levels in the treatment facilities are below Occupational Safety and Health Administration (OSHA) requirements.

Ozone systems, therefore, have the following major components (Figures 3-9 and 3-10):

- Air preparation (dryer) or oxygen supply and feed system
- Ozonator or ozone generator
- Contactor or eductor
- Ozone destruction unit
- Ozone residual monitors

Ozone feed (dosage) is controlled by adjusting the gas flow rate (for large changes in plant water flow) and the generator power. These adjustments cause a change in the ozone concentration in the feed to the contactors. This value combined with the water flow rate determines the applied ozone dosage. This is the key parameter for controlling ozone system performance.

Determining the optimum ozone dosage depends on several factors, including CT requirements, ozone residual target, transfer efficiency, ozone demand, oxidation objective, and ozone decay characteristics. Dosage calculations are included in chapter 4.

Chlorine Dioxide Systems Using Chlorine Gas

On-site generation is necessary because the compressed liquid is explosive at room temperature. It is an unstable gas and, therefore, is not stored or sent in bulk. It is most commonly produced by reacting chlorine with sodium chlorite. There are several generation system choices using chlorine solution, chlorine gas and chlorite solution, or solid chlorite (Figure 3-11). Chlorite solution spills should be neutralized with sodium sulfite. *Do not use sodium bisulfite.*

Regulations limit the applied dosage of chlorine dioxide (0.8 mg/L MCL) and the residual chlorite (1.0 mg/L MCL). Therefore, the dosage must be measured and controlled. It is important to measure the generator output of chlorine dioxide concentration. System calibration requires several generator settings to ensure accurate dosing. Measuring chlorine dioxide concentration after a predetermined contact time establishes the residual. This value verifies CT compliance (if it is used for primary disinfection) or is the control for the applied dosage.

Some chlorine dioxide feed systems use a batch generation process. Regardless of the generator used, it fills a batch storage tank. A feed pump meters the chemical to the feed location. Continuous

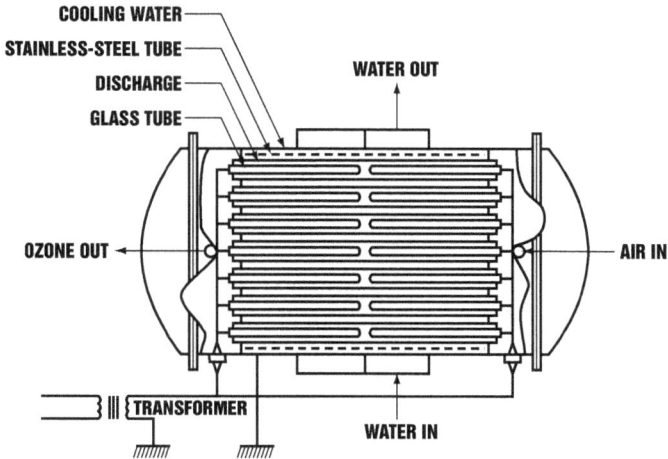

From Letterman, Raymond D., and American Water Works Association, Water Quality
and Treatment: A Handbook of Community Water Supplies. 5th ed. Copyright © 1999
McGraw-Hill. Reproduced with permission of The McGraw-Hill Companies.

Figure 3-9   Ozone generator

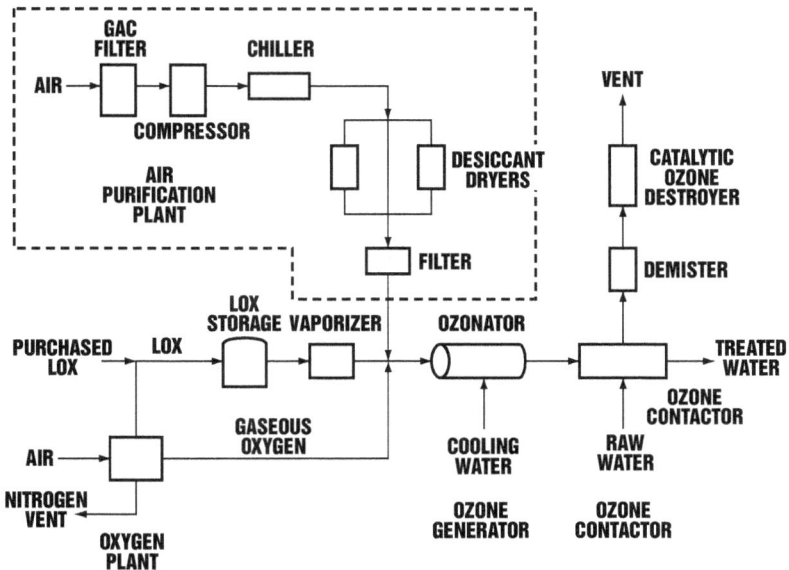

From Letterman, Raymond D., and American Water Works Association, Water Quality
and Treatment: A Handbook of Community Water Supplies. 5th ed. Copyright © 1999
McGraw-Hill. Reproduced with permission of The McGraw-Hill Companies.

Figure 3-10   Ozone system

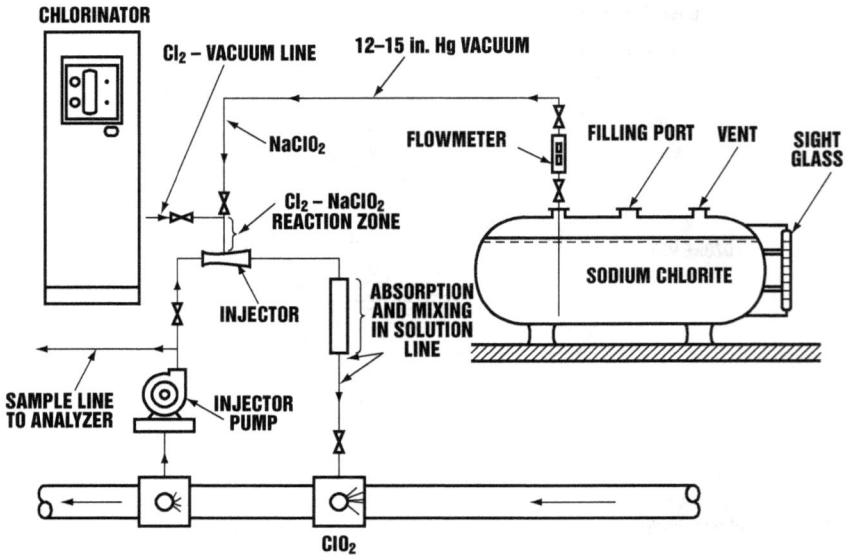

Figure 3-11   Chlorine dioxide generation system: Chlorine gas/chlorite solution

chemical analyzers measure the chlorine dioxide concentration from the generator and the batch tank, so the correct dosage matches the water flow. Also, chlorite is checked at several locations to ensure compliance with regulatory requirements.

Feeding Gases in Water Treatment

Only a few chemicals used in water treatment are gases. Many drinking water suppliers have switched from gases to liquids for safety and ease of operation. However, some chemicals are still used in a gas form. The chemical feed systems vary in complexity. Simple systems feed the gas directly to the application point and bubble it into the water without any intervening steps. Complex systems use both gases and liquids to produce the chemical, store batches at various strengths, and feed this material into a variable flow of water. Regardless of the degree of complexity, and using the proper procedures, feeding gases can be carried out accurately and safely.

Chapter 4

Chemical Feed of Gases— Calculations and Examples

Liquefied gases under pressure are commonly fed as gases when used in water treatment processes. Calculations are necessary to determine the correct settings to deliver the desired amount of chemical to the water flow. Also, it is important to monitor the chemical use to anticipate when to order the delivery of more supplies.

Consistent units of measure used in the calculation formulas (calculators) are a must. For example, each calculator is either in US or SI (metric) units (Appendix B). To use these calculators, it may be necessary to convert the units of measure to match those in the formula (Tables 4-1 and 4-2). The calculators here are simplified, often with the constants already incorporating the conversion factors. The derivation of the simplified calculators is given in Appendix D. Chemical feed calculators are provided for the most common treatment situations. Combinations of these calculators may be needed for other treatment systems.

Gas Feed Directly into the Water Flow

It is common to feed pure compressed gases such as chlorine, oxygen, sulfur dioxide, anhydrous ammonia, and carbon dioxide directly into the water flow. The following calculators are used to determine the feeder setting and daily use for this feed situation. Table 4-2 is used for conversions from lb/day or g/hr to lb/hr, g/day, or kg/day.

c4-1 $\boxed{\text{Feeder setting (lb/day)}} = 8.34 \times \boxed{\text{Dosage (mg/L)}} \times \boxed{\text{Water flow (mgd)}}$

c4-2 $\boxed{\text{Feeder setting (g/hr)}} = 41.7 \times \boxed{\text{Dosage (mg/L)}} \times \boxed{\text{Water flow (ML/day)}}$

Table 4-1   Flow conversions

US Units		SI Units		
gpm*	*mgd**	*m³/day**	*L/min**	*ML/day**
50	0.072	200	139	0.2
100	0.144	400	278	0.4
200	0.288	600	417	0.6
300	0.432	800	556	0.8
400	0.576	1,000	694	1
500	0.72	5,000	3,472	5
600	0.864	10,000	6,944	10
700	1.008	20,000	13,889	20
800	1.152	30,000	20,833	30
900	1.296	40,000	27,778	40
1,000	1.44	50,000	34,722	50
5,000	7.2	60,000	41,667	60
10,000	14.4	70,000	48,611	70
20,000	28.8	80,000	55,556	80
50,000	72	90,000	62,500	90
100,000	144	100,000	69,444	100
150,000	216	500,000	347,222	500
200,000	288	1,000,000	694,444	1,000

*   gpm = gallons per minute, mgd= million gallons per day, m³/day = cubic meters per day, L/min = liters per minute, ML/day = million liters per day.

Calculation Examples

Example 4-1   A water flow of 2 mgd is treated with a chlorine dosage of 1.5 mg/L. What is the gas chlorinator setting for this situation?

Use calculator c4-1.

| Feeder setting (lb/day) | = 8.34 × | Dosage (mg/L) | × | Water flow (mgd) |

$$\text{Feeder setting (lb/day)} = 8.34 \times 1.5 \text{ mg/L} \times 2 \text{ mgd}$$

$$= 25.02 \text{ lb/day}$$

Example 4-2   A water flow of 5 ML/day is treated with a carbon dioxide dosage of 10 mg/L/. What is the gas feeder setting for this situation?

Use calculator c4-2 (SI units).

| Feeder setting (g/hr) | = 41.7 × | Dosage (mg/L) | × | Water flow (ML/day) |

10 ↓   5 ↓

Feeder setting   = 41.7 ×   10 mg/L × 5 ML/day
(g/hr)

= **2,085 g/hr**

Example 4-3   A water flow of 100 gpm is treated with an anhydrous ammonia dosage of 0.5 mg/L. What is the feeder setting for this situation?

Use calculator c4-1 (US units).
100 gpm × 1,440 m/day × 1MG/1,000,000g = 0.144 MGD

| Feeder setting (lb/day) | = 8.34 × | Dosage (mg/L) | × | Water flow (mgd) |

0.5 ↓   0.144 ↓

Feeder setting   = 8.34 ×   0.5 mg/L × 0.144 mgd
(lb/day)

= **0.6 lb/day of ammonia**

Ops Tip

This calculator uses a flow in mgd so gpm must be converted to mgd first (Table 4-1).

Table 4-2   Feed rate conversions

US Units		SI Units		
lb/hr*	lb/day*	g/day*	kg/day*	g/hr*
0.002	0.05	24	0.024	1
0.04	1	48	0.048	2
0.08	2	72	0.072	3
0.13	3	96	0.096	4
0.17	4	120	0.12	5
0.21	5	144	0.144	6
0.42	10	168	0.168	7
1.04	25	192	0.192	8
2.08	50	216	0.216	9
3.13	75	240	0.24	10
4.17	100	1,200	1.2	50
10.42	250	2,400	2.4	100
20.83	500	12,000	12	500
41.67	1,000	24,000	24	1,000
83.33	2,000	120,000	120	5,000
125.00	3,000	240,000	240	10,000
208.33	5,000	1,200,000	1,200	50,000
416.67	10,000	2,400,000	2,400	100,000

*   lb/hr = pounds per hour, lb/day = pounds per day, g/day = grams per day, kg/day = kilograms per day, g/hr = grams per hour.

Ozone Feed Calculations

Ozone feed is similar to the pure gases (chlorine, sulfur dioxide, anhydrous ammonia, carbon dioxide), but it is complicated by the fact that the feed gas is not 100 percent pure ozone. Also, in ozone contactors, the transfer to the water is not 100 percent efficient. Rather, the feed gas is mostly dry air or dry oxygen that contains some ozone, depending on the generator operating conditions. The exact amount of ozone in the air or oxygen depends on the temperature, pressure, and feed gas density, as well as the voltage and other generator parameters.

The number of variables needed to determine precisely the ozone dosage prevents developing a single table. However, by allowing a small error (for example, within 1 to 2 percent) in the results, several variables can be fixed for most applications. Use the calculators (c4-3 through c4-7) to determine the ozone dose or the gas flow to the contactor for an oxygen feed-gas system.

Calculators c4-8 through c4-10 are for an air-feed ozone system. For a more rigorous calculation (or for calculations for pressure swing adsorption, vacuum swing adsorption, and vacuum-pressure swing adsorption systems), refer to *Ozone in Drinking Water Treatment* (Rakness 2005).

Oxygen Feed System Calculation Examples

Calculators c4-3 through c4-6 use a number of constants (as listed in Table D-1 in Appendix D). To use the calculators, insert the values for the percent weight ozone concentration, the gas flow to the ozone contact in standard cubic feet per minute (scfm) or cubic newton-meters per hour (Nm^3/hr), and the water flow rate.

c4-3 $\boxed{\text{Ozone dosage (mg/L)}} = 0.14 \times \boxed{\text{\%wt ozone concentration}} \times \boxed{\text{Gas flow (scfm)}} \div \boxed{\text{Water flow (mgd)}}$

c4-4 $\boxed{\text{Ozone dosage (mg/L)}} = 0.32 \times \boxed{\text{\%wt ozone concentration}} \times \boxed{\text{Gas flow (Nm}^3\text{/hr)}} \div \boxed{\text{Water flow (ML/day)}}$

c4-5 $\boxed{\text{Gas flow (scfm)}} = \boxed{\text{Ozone dosage (mg/L)}} \times \boxed{\text{Water flow (mgd)}} \div 0.14 \div \boxed{\text{\%wt ozone concentration}}$

c4-6 $\boxed{\text{Gas flow (Nm}^3\text{/hr)}} = \boxed{\text{Ozone dosage (mg/L)}} \times \boxed{\text{Water flow (ML/day)}} \div 0.32 \div \boxed{\text{\%wt ozone concentration}}$

Calculation Examples

Example 4-4   Calculate the ozone dosage for a pure oxygen system where the ozone concentration in the feed gas is 11%, the gas flow is 20 scfm, and the water flow is 10 mgd.

Use calculator c4-4.

Ozone dosage (mg/L)	= 0.14 ×	%wt ozone concentration	×	Gas flow (scfm)	÷	Water flow (mgd)

Dosage (mg/L)   = 0.14 × 11% × 20 scfm ÷ 10 mgd

= 3.08 mg/L ozone

Example 4-5   Determine the gas flow needed in a pure oxygen system to apply an ozone dosage of 1.5 mg/L to a water flow of 0.5 mgd where the ozone concentration in the feed gas is 10%wt.

Use calculator c4-5.

Gas flow (scfm)	=	Ozone dosage (mg/L)	×	Water flow (mgd)	÷ 0.14 ÷	%wt ozone concentration

Gas flow   = 1.5 mg/L × 0.5 mgd ÷ 0.14 ÷ 10%
(scfm)

= 0.54 scfm oxygen

Air Feed System Calculation Examples

Calculators c4-7 through c4-10 assume a number of constants (as listed in Table D-2 in Appendix D). To use the calculators, the values should be inserted for the percent weight ozone concentration, the gas flow to the ozone contact, and the water flow rate.

c4-7	Ozone dosage (mg/L)	= 0.13 ×	%wt ozone concentration	×	Gas flow (scfm)	÷	Water flow (mgd)

Ops Tip

Insert the %wt ozone concentration.

c4-8 $\boxed{\text{Ozone dosage (mg/L)}} = 0.29 \times \boxed{\text{\%wt ozone concentration}} \times \boxed{\text{Gas flow (Nm}^3\text{/hr)}} \div \boxed{\text{Water flow (ML/day)}}$

c4-9 $\boxed{\text{Gas flow (scfm)}} = \boxed{\text{Ozone dosage (mg/L)}} \times \boxed{\text{Water flow (mgd)}} \div 0.13 \div \boxed{\text{\%wt ozone concentration}}$

c4-10 $\boxed{\text{Gas flow (Nm}^3\text{/hr)}} = \boxed{\text{Ozone dosage (mg/L)}} \times \boxed{\text{Water flow (ML/day)}} \div 0.29 \div \boxed{\text{\%wt ozone concentration}}$

Calculation Examples

Example 4-6   Calculate the ozone dosage for an air feed system where the ozone concentration in the feed gas is 2%, the gas flow is 20 scfm, and the water flow is 10 mgd.

Use calculator c4-7.

$\boxed{\text{Ozone dosage (mg/L)}} = 0.13 \times \boxed{\underset{2}{\text{\%wt ozone concentration}}} \times \boxed{\underset{20}{\text{Gas flow (scfm)}}} \div \boxed{\underset{10}{\text{Water flow (mgd)}}}$

Dosage (mg/L) $= 0.13 \times 2\% \times 20 \text{ scfm} \div 10 \text{ mgd}$

= 0.52 mg/L ozone

Example 4-7   Determine the gas flow needed in a pure oxygen system to apply an ozone dosage of 1.5 mg/L to a water flow of 0.5 mgd where the ozone concentration in the feed gas is 3%wt.

Use calculator c4-9.

$\boxed{\text{Gas flow (scfm)}} = \boxed{\underset{1.5}{\text{Ozone dosage (mg/L)}}} \times \boxed{\underset{0.5}{\text{Water flow (mgd)}}} \div 0.13 \div \boxed{\underset{3}{\text{\%wt ozone concentration}}}$

Gas flow (scfm) $= 1.5 \text{ mg/L} \times 0.5 \text{ mgd} \div 0.13 \div 3\%$

= 1.92 scfm

Chlorine Dioxide Feed Calculations

Chlorine dioxide is produced on-site, as needed, for the application. At least two chemicals are combined to form chlorine dioxide. While a few proprietary chlorine dioxide feed systems exist, most systems used in water treatment plants combine chlorine gas (some use liquid sodium hypochlorite) with sodium chlorite (either solid or liquid). A chlorine dioxide solution is then fed at the point of application. Although chlorine dioxide feed systems are often a hybrid of gas and liquid (or solid), this discussion is included in the gas feed section of this book because the chlorine gas is commonly the controlling factor in this system.

It is important to follow equipment manufacturer's instructions to generate chlorine dioxide efficiently. Adjustments to chlorine and sodium chlorite feed may be necessary to optimize production and minimize forming unwanted by-products (chlorate, chlorite, or excess chlorine). The high-yield production of chlorine dioxide depends on the relationship of chlorine to sodium chlorite, pH, and other factors. Precise and accurate control of all variables is necessary to produce the highest-purity product and to minimize undesirable by-products.

Chlorine dioxide dosage control is often based on the residual measurement. Analyzers measure chlorine dioxide in the treated water and feed a control signal back to the generator if adjustment is needed. Some systems instead have analyzers on the generator effluent. In this case, the chlorine dioxide concentration produced by the generator is used to adjust the feed parameters.

Batch chlorine dioxide systems use ideal conditions to produce a high-purity solution that is stored in a tank and fed into the water by a chemical metering pump. The concentrated chlorine dioxide solution is rather stable, and storage of a few days is usually acceptable.

Several chlorine dioxide calculators and tables are illustrated in c4-11 through c4-18 and Table 4-3. These are useful to decide the generator production needed for a given plant flow and a specific dosage. Also, chlorine and sodium chlorite amounts are calculated for various plant flows and dosages. The calculators are based on chemical reactions and may not reflect results obtained from specific generators and application conditions. However, these may still be useful to more quickly set generator conditions and adjust settings to optimize system performance.

Table Tamer	To determine the chlorine dioxide feed for dosages other than 1 mg/L, multiply the table value by the dosage in mg/L.

Table 4-3   Chlorine dioxide feed (for 1 mg/L dosage)

US Units		SI Units	
Plant Flow, mgd	ClO_2, lb/day	Plant Flow, ML/day	ClO_2, kg/day
0.1	0.834	0.5	0.5
0.5	4.17	1	1
1	8.34	5	5
5	41.7	10	10
10	83.4	25	25
25	208.5	50	50
50	417	75	75
75	625.5	100	100
100	834	150	150
150	1,251	200	200
200	1,668	300	300
300	2,502	400	400

Chlorine Dioxide Feed Calculators

c4-11 ClO_2 feed (lb/day) $= 8.34 \times$ Dosage (mg/L) $\times$ Water flow (mgd)

c4-12 ClO_2 feed (kg/day) $=$ Dosage (mg/L) $\times$ Water flow (ML/day)

Example 4-8   What is the chlorine dioxide feed needed if the water flow is 2 mgd and the dosage is 0.5 mg/L?

You can use either Table 4-3 or calculator c4-11 for this problem.

• Using Table 4-3.

From the table, ClO_2 feed for 1 mgd flow and 1 mg/L dosage is 8.34 lb/day. The water flow for this problem is 2 mgd, so multiply by 2 = 16.68 lb/day. But the dosage is only 0.5 mg/L, which is ½ of the dosage = **8.34 lb/day.**

So for 2 mgd, multiply this by 2 = 16.68 lb/day. This is for 1 mg/L dosage. Multiply this value by 0.5 mg/L dosage = **8.34 lb/day.**

• Using calculator c4-11.

		0.5		2
ClO_2 feed (lb/day)	= 8.34 ×	Dosage (mg/L)	×	Water flow (mgd)

ClO_2 feed (lb/day)   = 8.34 × 0.5 mg/L × 2 mgd

= **8.34 lb/day**

Chlorine and Sodium Chlorite Feed Calculators

The feed rate of chlorine and sodium chlorite is linked to the feed of chlorine dioxide because these are the primary chemicals used in most generation systems. The most common sources of chlorine are compressed gas and liquid sodium hypochlorite (12 percent trade is a common solution strength). Sodium chlorite is commonly used as a solid (about 65 percent sodium chlorite) and as a 25 percent strength solution.

The chemical reaction of sodium chlorite ($NaClO_2$) and chlorine (Cl_2) to produce chlorine dioxide (ClO_2) and salt ($NaCl$) is given in the equation below.

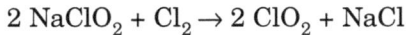

$$2\ NaClO_2 + Cl_2 \rightarrow 2\ ClO_2 + NaCl$$

This equation results in the following weight relationship, where g is equal to grams.

1.34 g $NaClO_2$ combine with 0.53 g Cl_2 to produce 1 g ClO_2

This relationship is shown in the calculator below.

c4-13

$$\boxed{\begin{array}{c}\text{Chlorine}\\(0.53 \text{ wt units/day})\end{array}} = \boxed{\begin{array}{c}\text{ClO}_2 \text{ production}\\(1 \text{ wt units/day})\end{array}}$$

and

$$\boxed{\begin{array}{c}\text{Sodium chlorite}\\(1.34 \text{ wt units/day})\end{array}} = \boxed{\begin{array}{c}\text{ClO}_2 \text{ production}\\(1 \text{ wt units/day})\end{array}}$$

Expressing the weight relationship in this way allows the sodium chlorite consumption to be calculated for any concentration or for any chlorine dioxide production amount. Also, any weight units can be used as long as all the units used are identical.

This relationship can be further expanded by substituting the equation for chlorine dioxide feed shown in calculators c4-12 and c4-13. Thus, we have the basic relationship (c4-14) that is useful in calculating the sodium chlorite of any concentration needed to produce a given amount of chlorine dioxide. Remember, these relationships are drawn from the chemical equations, thus 100 percent efficiency is assumed. Lower efficiency therefore results in amounts that differ from these calculated values. However, the calculated values are useful to estimate chemical use, and these can be used to evaluate generator efficiency (see "A Generator Purity Calculation," p. 53).

c4-14 *Sodium Chlorite Usage*

$$\boxed{\begin{array}{c}\text{Wt units}\\\text{of sodium}\\\text{chlorite for}\\\text{concentration}\\\text{used}\end{array}} = \boxed{\begin{array}{c}1.34 \text{ wt units/}\\\text{day/1 wt unit of}\\\text{ClO}_2 \text{ produced}\\\text{by pure sodium}\\\text{chlorite}\end{array}} \times \boxed{\begin{array}{c}\text{Number of}\\\text{wt units}\\\text{of ClO}_2\\\text{produced}\end{array}} \div \boxed{\begin{array}{c}\text{Decimal}\\\text{concentration}\\\text{of sodium}\\\text{chlorite}\end{array}}$$

Example 4-9   Calculate the amount of 25 percent sodium chlorite solution (theoretical) needed to generate 400 lb/day of chlorine dioxide. How many gallons of 25 percent sodium chlorite solution is this?

Use calculator c4-14.   400   0.25

| Wt units of sodium chlorite for concentration used | = | 1.34 wt units/ day/1 wt unit of ClO_2 produced by pure sodium chlorite | × | Number of wt units of ClO_2 produced | ÷ | Decimal concentration of sodium chlorite |

= 1.34 × 400 lb/day ÷ 0.25

= 2,144 lb/day of 25% sodium chlorite solution

To convert the lb/day to gal/day, you must know the specific gravity or density. The specific gravity of 25 percent sodium chlorite is 1.2. Convert specific gravity (has no units) to density (units of lb/gal) by using an 8.34 conversion factor as follows (chapter 8, p. 100):

Density lb/gal = specific gravity (1.2) × density of water (8.34 lb/gal) = 10 lb/gal

Now using this density, convert lb/day to gal/day

2,144 lb/day ÷ 10 lb/gal = 214.4 gal/day of sodium chlorite solution

The chlorine needed to produce a given amount of chlorine dioxide is similarly determined by calculator c4-15. Remember that this is based on the chemical equation that assumes 100 percent efficiency. Generators are not always this efficient, so the chlorine needed may be more than the result of this calculation. However, this calculated value is useful in estimating the amount needed and in determining the generator efficiency.

c4-15   *Chlorine Usage*

| Wt units of chlorine for concentration used | = | 0.53 wt units/ day/1 wt unit of ClO_2 consumption of pure chlorine | × | Number of wt units of ClO_2 consumption | ÷ | Decimal concentration of chlorine |

Example 4-10 Calculate the amount of 12 percent sodium hypochlorite (chlorine) solution (theoretical) needed to generate 50 lb/day of chlorine dioxide. How many gallons of 12 percent sodium hypochlorite is this?

Use calculator c4-15.

Wt units of chlorine for concentration used	=	0.53 wt units/ day/1 wt unit of ClO_2 consumption of pure chlorine	×	Number of wt units of ClO_2 consumption	÷	Decimal concentration of chlorine

$$= 0.53 \times 50 \text{ lb/day} \div 0.12$$

= 220.8 lb/day of 12% sodium hypochlorite

To convert the lb/day to gal/day, you must know the specific gravity or density. The specific gravity of 12 percent sodium hypochlorite is 1.2. Convert specific gravity (has no units) to density (units of lb/gal) by using an 8.34 conversion factor as follows (chapter 8, page 100):

Density lb/gal = specific gravity (1.2) × density of water (8.34 lb/gal) = 10 lb/gal

Now using this density, convert lb/day to gal/day.

220.8 lb/day ÷ 10 lb/gal = 22.1 gal/day of 12% sodium hypochlorite solution

Chlorine Gas/Sodium Chlorite Solid Generator Systems

These systems generate chlorine dioxide gas so the dosage calculations are the same as pure gas systems (c4-1 and c4-2). Chlorine dosage produces chlorine dioxide according to the theoretical relationship above (0.53 wt units of chlorine produce 1 wt unit of chlorine dioxide), but a maximum production concentration of about 8 percent.

Gas Feed Calculators

Theoretical chlorine feed setting for a chlorine dioxide production rate.

c4-1	Feeder setting (lb/day)	= 8.34 ×	Dosage (mg/L)	×	Water flow (mgd)

c4-2	Feeder setting (g/hr)	= 41.7 ×	Dosage (mg/L)	×	Water flow (ML/day)

Example 4-11 Calculate the chlorine feed setting for a chlorine dioxide dosage of 0.5 mg/L and a water flow of 0.3 ML/day.

Use calculator c4-2.   0.5   0.3

Feeder setting (g/hr)	= 41.7 ×	Dosage (mg/L)	×	Water flow (ML/day)

Chlorine dioxide, g/hr = 41.7 × 0.5 mg/L × 0.3 ML/day

= **6.3 g/hr**

From the chlorine/chlorine dioxide production relationship, 0.53 wt of chlorine produces 1 wt of chlorine dioxide, calculator c4-13.

Chlorine (0.53 wt units/day)	=	ClO_2 production (1 wt units/day)

6.3 g/hr chlorine dioxide × 0.53 = 3.3 g/hr chlorine

Chlorine Dioxide Concentration in Generator Effluent

The concentration of chlorine dioxide in the generator effluent is measured in some systems and is used as a dosage control. This value is calculated from the required chlorine dioxide production (calculators c4-12 and c4-13) and the water flow rate through the generator (for liquid sodium chlorite systems).

c4-16	Generator ClO_2 concentration (mg/L)	= 82.6 ×	ClO_2 production (lb/day)	÷	Water flow through the generator (gpm)

c4-17	Generator ClO_2 concentration (mg/L)	= 694.4 ×	ClO_2 production (kg/day)	×	Water flow through the generator (L/min)

The calculations for the ClO_2 production (consumption) can be substituted into calculators c4-12 and c4-13. The resultant calculators, c4-18 and c4-19, determine the concentration in the generator effluent that matches a specific water flow and dosage. Many systems use this value as a control, measuring the concentration of the generator effluent continuously.

c4-18 $\boxed{\begin{array}{c}\text{Generator} \\ \text{ClO}_2 \\ \text{concentration} \\ \text{(mg/L)}\end{array}} = 694.4 \times \boxed{\begin{array}{c}\text{Dosage} \\ \text{(mg/L)}\end{array}} \times \boxed{\begin{array}{c}\text{Water} \\ \text{flow} \\ \text{(mgd)}\end{array}} \div \boxed{\begin{array}{c}\text{Water flow} \\ \text{through the} \\ \text{generator} \\ \text{(gpm)}\end{array}}$

c4-19 $\boxed{\begin{array}{c}\text{Generator} \\ \text{ClO}_2 \\ \text{concentration} \\ \text{(mg/L)}\end{array}} = 694.4 \times \boxed{\begin{array}{c}\text{Dosage} \\ \text{(mg/L)}\end{array}} \times \boxed{\begin{array}{c}\text{Water} \\ \text{flow} \\ \text{(ML/} \\ \text{day)}\end{array}} \div \boxed{\begin{array}{c}\text{Water flow} \\ \text{through} \\ \text{the} \\ \text{generator} \\ \text{(L/min)}\end{array}}$

Example 4-12 What is the generator effluent concentration needed for a 18 mgd plant flow, a ClO_2 dosage of 0.4 mg/L, and where the water flow through the generator is 12 gpm?

Use calculator c4-18.   0.4   18   12

$\boxed{\begin{array}{c}\text{Generator} \\ \text{ClO}_2 \\ \text{concentration} \\ \text{(mg/L)}\end{array}} = 694.4 \times \boxed{\begin{array}{c}\text{Dosage} \\ \text{(mg/L)}\end{array}} \times \boxed{\begin{array}{c}\text{Water flow} \\ \text{(mgd)}\end{array}} \div \boxed{\begin{array}{c}\text{Water flow} \\ \text{through the} \\ \text{generator} \\ \text{(gpm)}\end{array}}$

ClO_2 concentration in generator effluent = 694.4 × 0.4 mg/L × 18 mgd ÷ 12 gpm

= 416.6 mg/L

A Generator Purity Calculation

There are several ways to calculate generator yield (efficiency). One calculation method defines yield as the ratio of chlorine dioxide concentration to chlorine dioxide concentration plus concentrations of chlorite and chlorate. Purity is another related calculation that includes excess chlorine in the ratio. The theoretical relationship shown in c4-14 can be used as another possible way of determining this value. The chlorine dioxide value from this relationship assumes 100 percent efficiency of the generator. This amount can be compared to the measured chlorine dioxide concentration and the purity thus determined.

c4-20   Purity (%) = $\boxed{\begin{array}{c}\text{Measured ClO}_2 \\ \text{concentration of} \\ \text{the generator}\end{array}} \div \boxed{\begin{array}{c}\text{Theoretical ClO}_2 \\ \text{concentration} \\ \text{from c4-17}\end{array}} \times 100$

Example 4-13 What is the purity of the generator effluent described in Example 4-12 if the measured chlorine dioxide concentration of the effluent is 400 mg/L?

Use calculator c4-20

c4-20   Purity (%) = $\dfrac{\text{Measured } ClO_2 \text{ concentration of the generator}}{400} \div \dfrac{\text{Theoretical } ClO_2 \text{ concentration from c4-18}}{416.6} \times 100$

From Example 4-12 and calculator c4-18, theoretical ClO_2 is 416.6 mg/L

Purity = 400 mg/L ÷ 416.6 mg/L × 100 = **96.0%**

The remaining 4.0% is chlorite, chlorate, and excess chlorine.

Generator Efficiency (Yield)

Amperometric titration (*Standard Methods for the Examination of Water and Wastewater* [Eaton and Franson 2005], method 4500) of the generator effluent is used to determine the concentration of the various products of the reaction of chlorine with sodium chlorite. This test method determines the concentration of chlorine dioxide, chlorite, chlorate, and excess chlorine. The ratio of chlorine dioxide to the total of all chemical species is the efficiency.

c4-21   Efficiency (%) = 100 × $\dfrac{\text{Measured } ClO_2 \text{ concentration of the generator}}{} \div \dfrac{\text{Sum of } ClO_2 \text{ concentration + chlorite concentration + (0.81 × chlorate concentration)}}{}$

Chapter 5

Dry Chemical Feeders and Systems

Many chemicals used in water treatment are delivered and fed in dry form. Table 5-1 lists some common dry treatment chemicals and their common forms. Chemical feed systems measure these chemicals either by volume (volumetric) or by weight (gravimetric). Dry chemicals are sometimes delivered directly to the water. It is far more common, however, to mix the dry chemical with a small volume of water first. The concentrated solution or *slurry* is then fed into the water, where it is thoroughly mixed. This method is better at dispersing the chemical into the treated water.

Most dry chemical feed systems have several major parts. Dry chemicals are delivered in bulk or packaged in bags or drums. Enough inventory is stored on-site to ensure an uninterrupted supply. Chemicals are then loaded into hoppers that supply the feeder. In some smaller treatment plants, the bag or drum is used as the hopper. The material is metered from the feeder into a solution tank, where it is mixed with water. The solution is then dispersed into the water stream (sometimes the solution is held in a day tank first), where it is thoroughly mixed. A typical dry chemical feed system is shown in Figure 5-1.

Lime for pH adjustment and softening is a major dry chemical used in water treatment. The two forms commonly used are hydrated lime ($Ca[OH]_2$) and quicklime (CaO). Quicklime is slaked with a small volume of water (converted into hydrated lime) to become a slurry that is delivered to the water. Quicklime reacting with water produces much heat. Operators must take precautions to avoid burns when working with quicklime. Figure 5-2 shows a typical quicklime feed system.

Table 5-1   Common dry chemicals used in water treatment

Chemical, Formula, Common Name	Grade or Available Forms	Shipping Containers	Approximate Solubility in Water (gm/100 ML) at 20°C	Commercial Strength, %
Activated carbon, PAC, GAC	Powder, granules	Bags: 35 lb; drums: 5, 25 lb; totes: bulk	Insoluble—slurry	10–90
Aluminum sulfate, Al_2O_3, dry alum	Lump, granules, ground, powder	Bags: 100 lb; drums: 25–250 lb; bulk	87.3	98 or 17
Calcium hydroxide, $Ca(OH)_2$, hydrated lime	Light powder, powder	Bags: 50–100 lb; bulk	0.16	82–95
Calcium oxide, CaO, quicklime	Pebble, lump, ground, granules, crushed	Bags: 100 lb; bulk	Reacts with water to form hydrated lime, solubility 0.16	70–96
Copper sulfate, $CuSO_4$	Lump, crystal, powder	Bags: 100 lb; drums	26.3	99
Ferric chloride, $FeCl_3$, iron chloride	Lumps, granules	Keg: 100–450 lb; drums: 150–630 lb	91.2	60–97
Ferric sulfate, $Fe_2(SO_4)_3$	Granules	Bags: 100 lb; drums: 400 lb; bulk	Soluble	68–76
Ferrous sulfate, $FeSO_4$	Granules, crystals, powder, lumps	Bags: 100 lb; totes: 400 lb; bulk	48.5	55
Potassium permanganate, $KMnO_4$	Crystal	Bags and drums: 25–600 lb; bulk	5.0	97–99

Table continued next page

Table 5-1   Common dry chemicals used in water treatment   (continued)

Chemical, Formula, Common Name	Grade or Available Forms	Shipping Containers	Approximate Solubility in Water (gm/100 ML) at 20°C	Commercial Strength, %
Sodium bicarbonate, $NaHCO_3$, baking soda	Powder, granules	Bags: 100 lb; drums: 25–400 lb	9.6	99
Sodium carbonate, Na_2CO_3, soda ash	Granules, powder	Bags: 100 lb; drums: 25–100 lb; bulk	21.5	99.2
Sodium chloride, NaCl, salt	Rock, powder, crystal, granules	Bags: 100 lb; drums; bulk	36.0	98
Sodium hexametaphosphate, $(NaPO_4)_6$	Glass, powder, flake	Bags: 100 lb; drums: 100–320 lb	Soluble	67
Sodium phosphate, NaH_2PO_4, monophosphate	Powder, crystal	Bags: 100 lb; drums: 125 lb; barrel: 320 lb	85–98	97–98
Sodium silicofluoride, Na_2SiF_6	Granules, powder	Bags: 100 lb; drums: 25–375 lb	0.7	98.5

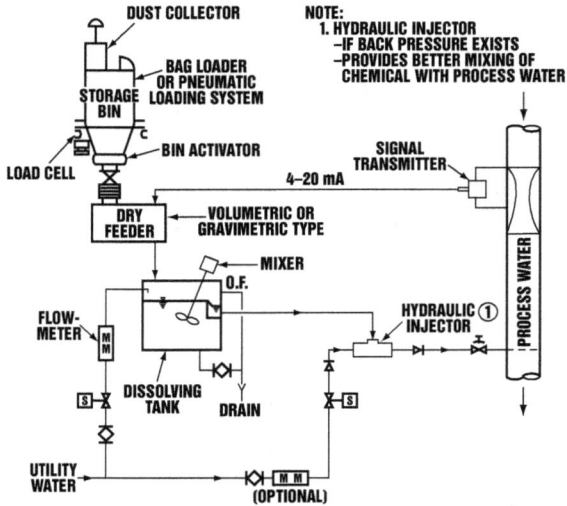

From Kawamura, Susumu, Integrated Design and Operation of Water Treatment Facilities, 2nd ed. Copyright © 2000 John Wiley & Sons, Inc. Reprinted with permission of John Wiley & Sons, Inc.

Figure 5-1   Typical dry chemical feed system

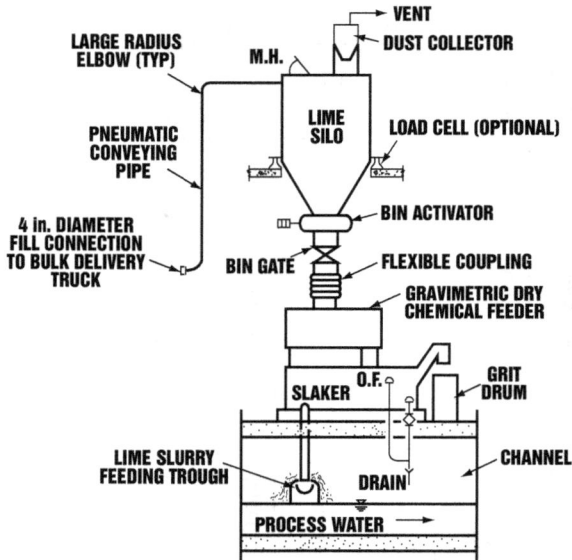

From Kawamura, Susumu, Integrated Design and Operation of Water Treatment Facilities, 2nd ed. Copyright © 2000 John Wiley & Sons, Inc. Reprinted with permission of John Wiley & Sons, Inc.

Figure 5-2   Typical quicklime feed system

Chemical Storage and Handling

Dry chemicals arrive at the treatment plant in bags, drums, bulk delivery trucks, or railcars. Often bags and drums are on pallets, and these are moved to storage areas within the plant. Lifting chemical bags or other containers is one of the most common causes of injuries. Operators should use proper lifting methods and hoists to avoid this hazard.

Bulk deliveries of dry chemicals are transferred to storage tanks by air pressure systems or conveyors. Drums used in smaller systems can be lifted and inverted with a drum inverter. Totes (or Super-bags) can be hoisted with a stationary lift or forklift. Powdered, granular, or even pebble forms of chemicals are transferred by air pressure systems. These systems have dust control equipment to reduce airborne chemical produced by this procedure. Sometimes the chemical is compacted in the receiving storage tank either by the transfer or by lengthy storage. Some storage tanks are fitted with vibration equipment to ensure the material can be efficiently removed. Air jets within the tanks are also used to loosen the material. Bulk chemical storage tanks (Figure 5-3) and transfer equipment need periodic maintenance for proper operation.

Figure 5-3   Dry chemical storage tank

Courtesy of Mike Barsotti

Figure 5-4   Permanganate drum hopper

Chemical Feeders

Both volumetric and gravimetric feed equipment are used to feed dry chemicals. Volumetric feeders (Figure 5-6) commonly use a rotating screw to meter the chemical from the supply hopper to a dissolving tank. Varying the rotation speed changes the chemical feed rate. This rotation speed may be paced with flow rate of the water being treated. The accuracy of the feeder is affected by the bulk density of the material and the size of the rotating screw relative to the amount of chemical fed. For best feed results, the chemical should have a uniform consistency. Some chemicals are manufactured in free-flowing grade. Many chemicals satisfy this requirement, and thus, volumetric feeders can be accurate.

Gravimetric feeders (Figure 5-6) may also use a rotating screw to move the material from the supply hopper to a belt conveyor. The belt is fitted with a weight cell so the material is weighed as it is fed. A control signal continuously adjusts the rotating screw based on the weight of the material on the belt. Gravimetric feeders give a more direct measure of the chemical being fed by the equipment than do volumetric feeders, which must convert volume to weight (one way

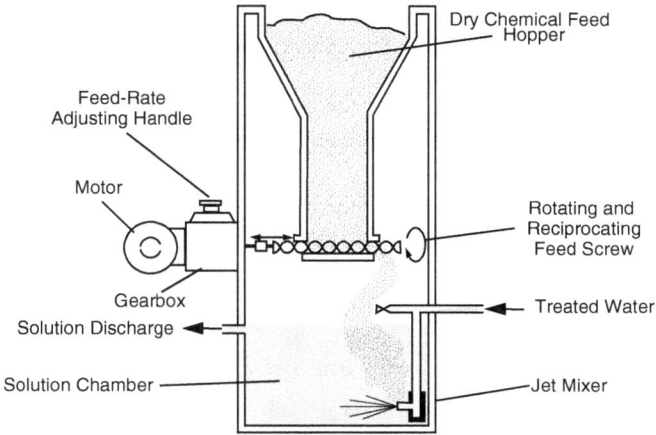

Figure 5-5   **Volumetric feeder**

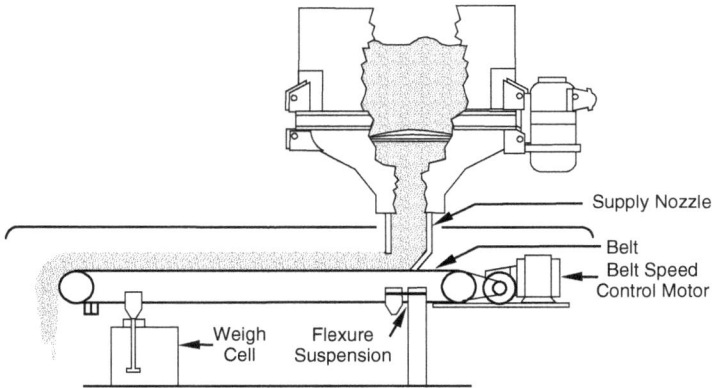

Figure 5-6   **Gravimetric feeder**

around this is to use a scale under the hopper to measure weight loss over time).

Feeder Calibration

Gravimetric and volumetric feeders need periodic calibration. Even feeders with direct weight readout must be calibrated to ensure accurate feed is continuously maintained. The feeder, age, maintenance history, and operational experience influence the frequency of calibration. Volumetric feeders must be calibrated when a chemical batch is changed. This is because of a possible change in the consistency or bulk density of the material. Ancillary equipment (such as tank vibrators or air pulse systems) may cause changes to the feed equipment or the chemical being fed. These changes may force more frequent calibration. Feed equipment should be calibrated no less often than recommended by the manufacturer.

Dry chemical feeders are calibrated by weighing the chemical delivered from the feeder for a measured time period. An example of a calibration for a volumetric feeder is described in Example 5-1.

Dry Chemical Feeder Calibration Procedure

1. Record feeder setting.

2. Catch chemical from feeder and record the time (sec or min).

3. Weigh the chemical collected in step 2.

4. Repeat this for several feeder settings.

5. Plot the results (include the date and operator name).

6. Compare each calibration with previous calibrations (this may reveal some problem with the equipment).

c5-1   Feed rate calculation for feeder calibration (US units)

Feed rate (lb/hr)	= 60 ×	Weight of chemical (lb)	÷	Collection time (min)

c5-2   Feed rate calculation for feeder calibration (SI units)

Feed rate (kg/hr)	= 60 ×	Weight of chemical (kg)	÷	Collection time (min)

The feed rate used for equipment calibration is for the chemical as delivered, that is, the amount of bulk chemical delivered by the equipment in a measured time period. Chemical purity is not part of the calibration used in this procedure. The strength of the chemical is used when determining dosage calculations.

Calibration Devices

Dry chemical feeders, whether volumetric or gravimetric, require similar devices for calibration. A timer, a container to collect the chemical, and a scale are needed. The timer should count in seconds. The container must be large enough to collect the chemical over the needed duration and fit in the feeder or the chemical feed throat.

Example 5-1   Dry Chemical Feeder Calibration

A dry chemical feeder is running and the operator takes several calibration measurements. The first feeder setting is 100 on the machine. A weighted container is used to collect the chemical for exactly 2 min. The amount collected is weighted on the laboratory scale (2,633 grams). This amount must be converted to pounds (divide the grams weighed by the conversion factor of 454 g/lb) because the feed rate should be in lb/hr. The next setting is 200, and the chemical is collected for exactly 1 min. The amount collected is 2,315 g. This process is continued as shown in Table 5-2. The feed rate is calculated for each measurement according to the calculator c5-1, or c5-2 for SI units. (It should be noted that c5-2 uses kilograms, so grams must be converted to kilograms for use in the calculator.) Then the feed rates and feeder settings are plotted on a chart (Figure 5-7). This chart is dated and used until the machine is calibrated again.

Table 5-2   Calibration results (example 5-1)

Feeder Setting	Sample time (min)	Amount collected (g)	Amount collected (lb)	Feed rate (lb/hr)
0	0	0	0	0
100	2	2,633	5.8	174
200	1	2,315	5.1	306
300	1	3,314	7.3	438
400	0.5	2,179	4.8	576
500	0.5	2,588	5.7	684
600	0.5	3,087	6.8	816
800	0.5	4,116	9.1	1,092

The longer the collection duration, the more accurate the calibration. Typically, calibration collection time should be several minutes. The scale must then have adequate capacity to weigh the empty and full container. Scale accuracy must be at least 10 percent of the empty weight of the container. For example, if the container weighs 2 lb, the scale must be accurate to at least 0.2 lb or less.

Ops Tip   Calibrate dry chemical feeders often to ensure accuracy. Volumetric feeders may need to be calibrated when new chemical deliveries are received.

DRY CHEMICAL CALIBRATION CHART

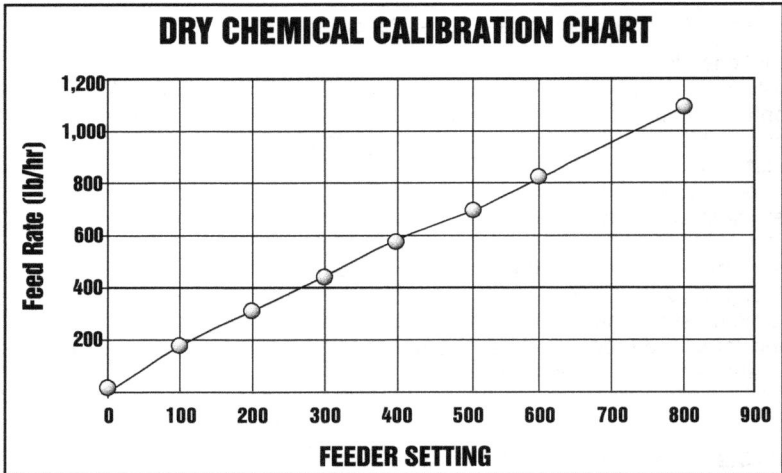

Figure 5-7   Calibration chart (example 5-1)

Containers to collect the chemical over a measured time period must be resistant to the chemical. Plastic containers are often used. However, other materials (metal or even cardboard) can serve this purpose, depending on the chemical. The collection container should be lightweight so it is easy to handle and does not interfere with the weight measurement of the chemical.

Solution Systems

Dry chemical solution systems are hybrids combining dry feeders with liquid solution (or dissolving) delivery units. Figures 5-1, 5-2, and 5-8 show some more common dry chemical solution system schematics. These systems include a dry chemical feeder, a solution tank or dissolving tank, and a conveyance device to move the liquid solution to the point of application.

Quicklime slaker systems (Figure 5-2) require special handling. Hydrated lime is first formed by slaking quicklime with a small amount of water. When water is added to quicklime, it produces heat, which helps dissolve (convert) the unreacted quicklime more efficiently to hydrated lime. The hydrated lime formed is mixed with more water to form a slurry that is delivered to the point of application. Huge lime softening plants in the midwestern United States use quicklime and this particular chemical feed system.

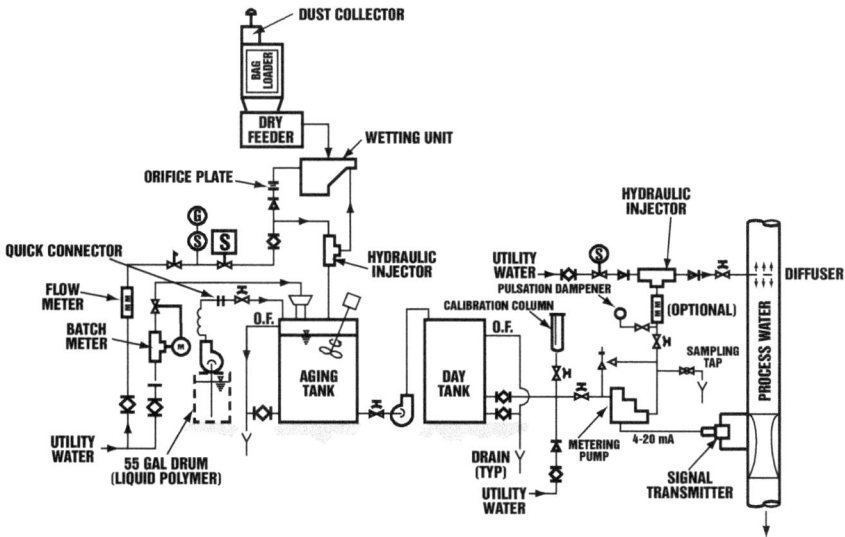

From Kawamura, Susumu, Integrated Design and Operation of Water Treatment Facilities, 2nd ed. Copyright © 2000 John Wiley & Sons, Inc. Reprinted with permission of John Wiley & Sons, Inc.

Figure 5-8   Dry chemical (polymer) solution system

Dry chemical solution systems use dry chemical feeders to meter the chemical. The solution part of the system is used to predissolve the chemical or to help disperse the chemical more efficiently into the process water stream. The chemical is measured by the dry chemical feeder. The flow of water through the solution part of the system does not affect the amount fed because the entire chemical from the dry feeder is delivered to the application point. The flow of solution water and the size of the dissolving tank do, however, affect the time needed to deliver the chemical from the feeder to the application point. This delay can be problematic for control systems because changes in dosage are not instantaneous. It may take many minutes or hours to see the result of a chemical change. In addition, insoluble dry chemicals, such as activated carbon, may develop clogs in long pipelines from the solution tank to injection point.

Ops Tip
Position feed equipment close to application point to reduce chemical addition delay.

Chapter 6

Dry Chemical Feed— Calculations and Examples

Many chemicals used in water treatment are fed dry (Table 1-1). Some can be fed directly into the process water stream where they are vigorously mixed. More commonly, dry chemicals are first mixed with some water (dissolved or as a slurry), and then the chemical solution is fed to the process water stream. Dry chemicals are also used to prepare batches of chemical solution. This solution is metered into the process water by liquid chemical metering pumps.

Liquid chemical feed calculations are the subject of chapter 8. This chapter includes the calculations for dry chemical feed either directly into process water or first into dissolving or slurry tanks, and to prepare solution batches. All dry chemical feeders are calibrated as needed to ensure accurate delivery of the chemical.

Full-Strength Chemical Feed Calculations

Dry chemicals are sometimes fed directly into a well-mixed process water stream. More often they are mixed with water in a dissolving tank or to prepare a slurry. The concentrated solution or slurry is then fed into the process water stream (see Figure 5-8). Calculating the feed rate for these systems is identical. The dry chemicals are considered full-strength (100 percent of the chemical being fed). This approach is useful when reviewing chemical inventory to make reorder decisions. Some operators still calculate the feed rate for dry polymers by adjusting for the purity (percent solids) of the dry material. It is more common, however, to consider whatever is delivered in the bag is pure polymer, and this would apply to the calculations in this section.

Ops Tip

Many dry chemicals are considered "full-strength" when used to prepare solutions.

c6-1	Feed rate (lb/day)	= 8.34 ×	Dosage (mg/L)	×	Water flow (mgd)

c6-2	Feed rate (g/hr)	= 41.7 ×	Dosage (mg/L)	×	Water flow (ML/day)

Calculators c6-1 and c6-2 can be used for many dry chemical feed situations. If the flow measurement or feed rate setting is in different units, the units should be converted in these calculators. Conversion formulas are provided for some of the more common units. For example, to use calculator c6-1 when the flow is in gallons per minute, the units can be converted to million gallons per day (mgd), and the value plugged in. The result will be a feed rate in pounds per day (lb/day). If the feed rate is in pounds per hour (lb/hr), the following formula can be used.

Example 6-1   Convert gpm to mil gal/day (mgd)

200 gpm is how many mgd?

gpm flow	× 0.00144 =	mgd flow
200 gpm	× 0.00144 =	0.288 MGD

Example 6-2   Convert lb/day to lb/hr

The feed rate water calculated at 300 lb/day. The chemical feeder setting is in lb/hr. What is the feeder setting?

lb/day	÷ 24 =	lb/hr
300 lb/day	÷ 24 =	12.5 lb/hr

Convert m^3/day to ML/day

m^3/day	× 1,000 =	ML/day

Convert kg/day to g/hr

kg/day	× 41.7 =	g/hr

Percent Active Ingredient Chemical Feed Calculations

Some dry chemicals contain varying amounts of the active ingredient for chemical application. The chemical feed calculations then require an adjustment for the amount of the active ingredient available in the dry chemical. Examples of this type of chemical are chemicals used for fluoridation (where the chemical is dosed to deliver a fluoride amount), calcium hypochlorite (to feed chlorine), phosphate chemicals (used to deliver a phosphate dosage), or copper sulfate (to deliver a specific copper dosage).

The percent active ingredient is often provided by the chemical manufacturer. This information is needed in some cases because the dry chemical may contain unknown amounts of inert substances needed to prevent caking or other storage and feed problems. If the complete dry chemical formula is known, the percent active ingredient can be calculated from the formula using the atomic weights of the elements (Appendix E).

Examples: sodium fluoride   NaF

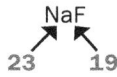

23   19

So 23 + 19 = 42

% F = 19/42 × 100 = **45.25%**

sodium fluorosilicate Na_2SiF_6

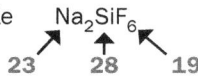

23   28   19

So (23 × 2) + 28 + (19 × 6) = 188

%F = (19 × 6)/188 × 100 = **60.7%**

calcium hypochlorite $Ca(OCl)_2$

40   16   35.5

So 40 + 2(16 + 35.5) = 142.6

% OCl = 2(16 + 35.5)/142.6 × 100 = **72%**

Ops Tip

Most commercial granular calcium hypochlorite is 65% available chlorine because the chemical is 90% pure.

Chemical feed calculations for dry chemicals where the percent active ingredient or purity is not 100 percent are based on the dosage of the active ingredient or desired chemical (c6-3 and c6-4).

c6-3 | Feed rate (lb/day) | = 834 × | Dosage (mg/L) | × | Water flow (mgd) | ÷ | % available active ingredient |

c6-4 | Feed rate (g/hr) | = 4,170 × | Dosage (mg/L) | × | Water flow (ML/day) | ÷ | % available active ingredient |

Example 6-3   What is the feed rate of sodium fluoride (lb/day) needed (45.25% available fluoride) to deliver a dosage of 0.8 mg/L fluoride to a flow of 5 mgd?

Use calculator c6-3.   **0.8**   **5**   **45.25**

| Feed rate (lb/day) | = 834 × | Dosage (mg/L) | × | Water flow (mgd) | ÷ | % available active ingredient |

= 834 × 0.8 mg/L × 5 mgd ÷ 45.25

= 73.7 lb/day of sodium fluoride

Feed rate results in lb/day or g/hr can be converted to lb/hr, kg/day, or m^3/day using the conversion factors given in the previous section on full-strength dry chemical feed calculations. Other conversion factors are provided in Appendix B.

Solution Preparation Using Dry Chemicals: Batch Preparation Calculations

Dry chemicals are often used to prepare solutions that are then fed into the process stream. The volume and the strength of the solution determine the amount of dry chemical needed. Calculators c6-5 and c6-6 are used to calculate the amount of dry chemical needed for any strength solution and for any volume. Tables 6-1 and 6-2 list the results for common volumes and solution strengths.

| c6-5 | lb of dry chemical needed | = 0.0834 × | Number of gallons to be prepared | × | % solution strength needed |

| c6-6 | kg of dry chemical needed | = 0.01 × | Number of liters to be prepared | × | % solution strength needed |

Table Tamer   8.34 lb of dry chemical are needed to prepare 100 gal of a 1% solution.

Table 6-1   Dry chemical amount (lb) to prepare solutions

Gallons of Solution	0.1% Strength	1% Strength	5% Strength	10% Strength	20% Strength
10	0.0834	0.834	4.17	8.34	16.68
20	0.1668	1.668	8.34	16.68	33.36
50	0.417	4.17	20.85	41.7	83.4
100	0.834	8.34	41.7	83.4	166.8
200	1.668	16.68	83.4	166.8	333.6
500	4.17	41.7	208.5	417	834
1,000	8.34	83.4	417	834	1,668
2,000	16.68	166.8	834	1,668	3,336

Table 6-2   Dry chemical amount (kg) to prepare solutions

Liters of Solution	0.1% Strength	1% Strength	5% Strength	10% Strength	20% Strength
50	0.05	0.5	2.5	5	10
100	0.1	1	5	10	20
200	0.2	2	10	20	40
500	0.5	5	25	50	100
1,000	1	10	50	100	200
2,000	2	20	100	200	400
5,000	5	50	250	500	1,000
10,000	10	100	500	1,000	2,000

Chemical Usage and Inventory Control

Estimating the anticipated chemical usage is necessary to ensure that adequate supplies are available to meet demands. Water treatment plant managers must maintain adequate chemical supplies without storing an excess. Storing excess chemicals can require unnecessary tankage or use valuable floor space. Excess chemicals stored for long periods can also lose strength, and disposal of outdated chemicals can increase costs.

Choosing the best time and amount to reorder is dependent on

- The predicted rate of chemical usage,
- The amount of chemical already in storage,
- The chemical delivery method and packaging,
- The shelf life of the chemical,
- The size and capacity for on-site storage,
- Accommodation for weekends and holidays (and seasonal issues),
- Delivery time from date the order is placed,
- Resource availability to receive chemicals upon delivery,
- Budget considerations.

The goal should be to maintain an uninterrupted supply of chemicals needed to treat the water without excess stored on-site. Many treatment plants require a minimum of a 15-day supply (Table 6-3) in storage at all times. Remotely located facilities or where delivery time is very long may require more on-site storage. For most plants, the following estimate may be used to determine when to reorder a chemical.

Reorder inventory trigger (minimum inventory):

Estimated daily use (lb/day) × (15 days + delivery time days + holiday or weekend days).

- Order an amount to get "full-truck" discount.
- Make sure there is adequate storage to accept full load.
- Make sure personnel is available to receive the load (quality control tests need to be performed) when it arrives.
- Make sure that any equipment needed to receive the load is available and functioning.

Example 6-4   The average quicklime dose used at a softening plant is 100 mg/L, and the average flow is 10 mgd. The plant has three storage silos each holding 50 tons. One of the silos is empty, and the other two are full. The normal delivery time is estimated at 5 days, but because there is an intervening holiday, it will take one extra day. Calculate the ordering trigger for this plant.

Quicklime usage

Calculate using c6-1

lb/day = 8.34 × 100 mg/L × 10 mgd = 8,340 lb of quicklime per day

Using the trigger formula

Trigger amount = 8,340 lb/day × (15-day minimum + 5 delivery days + 1 holiday)

= 175,140 lb

or 175,140 lb divided by 2,000 lb/ton = 87.6 tons

Currently, two of the three 50-ton silos are full, so 100 tons are on-site, and there is no need to order immediately. When should the order be placed?

Figure out how much quicklime can be used before reaching the minimum trigger:

100 tons on-site – 87.6-ton minimum inventory = 12.4 tons used before reorder.

Convert tons to pounds.

12.4 tons × 2,000 lb/ton = 24,800 lb before reorder.

Calculate how many days this will last.

24,800 lb divided by 8,340 lb used per day = 2.9 days.

In approximately three days, the minimum inventory trigger will be reached, and the order should be placed.

Table 6-3   Reorder trigger inventory amount—days of inventory for amount in pounds (lb) or tons

Chemical Usage (lb/day)	15-day Supply		20-day Supply		25-day Supply		30-day Supply	
	(lb)	(tons)	(lb)	(tons)	(lb)	(tons)	(lb)	(tons)
100	1,500	0.75	2,000	1	2,500	1.25	3,000	1.5
500	7,500	3.75	10,000	5	12,500	6.25	15,000	7.5
1,000	15,000	7.5	20,000	10	25,000	12.5	30,000	15
5,000	75,000	37.5	100,000	50	125,000	62.5	150,000	75
10,000	150,000	75	200,000	100	250,000	125	300,000	150
20,000	300,000	150	400,000	200	500,000	250	600,000	300

Table Tamer   If daily chemical usage is 1,000 lb, then a 20-day supply is 10 tons.

Chapter 7

Liquid Chemical Feed Equipment and Systems

Many water treatment operators consider liquid chemicals easy to feed and handle. Chemical suppliers deliver the chemicals in sealed containers, and the feed and delivery systems are contained. Accurate feed is assured when using properly sized, calibrated, and maintained chemical metering pumps and control systems. Liquids also mix easily into the process water stream. Table 7-1 lists some of the more common liquid chemicals used in water treatment. In addition to these, there are many liquid polyelectrolytes sometimes used in coagulation and filtration. They are not listed in the table because their properties vary depending on the specific product.

Table 7-1   Common liquid chemicals used in water treatment

Chemical, Formula, Common Name	Typical Strength (other strengths may be available)	Specific Gravity (approximate)	Density of Solution (lb/gal)
Aluminum sulfate, $Al_2(SO_4)_3$, alum	49%	1.33	11.2
Ammonium hydroxide, NH_4OH, aqua ammonia	30% NH_3	0.897	7.48
Ferric chloride, $FeCl_3$	40%	1.44	11.9
Hydrofluosilicic acid, H_2SiF_6	23%	1.21	10.1
Phosphoric acid, H_3PO_4	75%	1.57	13.1
Sodium hydroxide, NaOH, caustic soda	50%	1.54	12.8
Sodium hypochlorite, NaOCl, commercial bleach	12.5%	1.21	10.1
Sodium chlorite, $NaClO_2$	25%	1.23	10.25

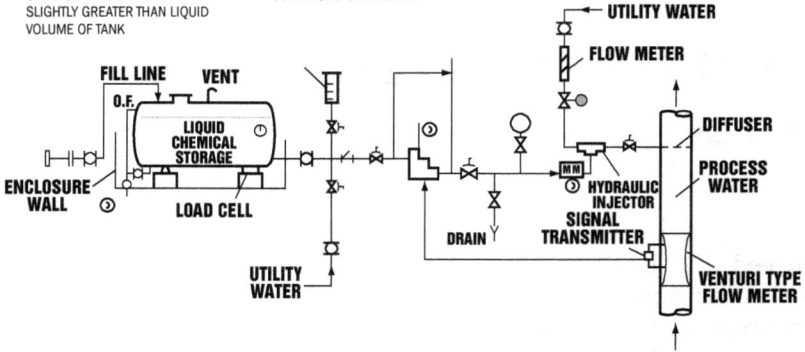

From *Kawamura, Susumu, Integrated Design and Operation of Water Treatment Facilities, 2nd ed. Copyright © 2000 John Wiley & Sons, Inc. Reprinted with permission of John Wiley & Sons, Inc.*

Figure 7-1   Typical liquid chemical feed system

Liquid chemical feed systems consist of storage tanks (with containment for hazardous chemicals), chemical piping, and metering pumps (with calibration equipment). Control systems monitor the level of chemical in the tanks, pressure and flow in the piping, and pump settings. A schematic of a typical liquid chemical feed system is shown in Figure 7-1.

Chemical Receiving and Storage

Liquid chemicals are delivered in drums, totes, tank trucks, and railcars. These containers are compatible with the chemical and are designed to hold the material for an extended time. The chemicals can be stored in the delivery containers or transferred to on-site bulk storage tanks. Before transferring chemicals to storage tanks or accepting delivery containers, sample and test the chemical, and verify that the proper chain of custody was followed from the manufacturer to the storage tank. The chemical must pass the receiving screening tests prior to acceptance.

Some chemicals can change when stored. Satisfying the storage conditions and restrictions for each chemical is critical. Some chemicals require periodic testing to ensure accurate chemical feed because the strength of the chemical may change. Other chemicals

Table 7-2   Liquid water treatment chemicals—special storage considerations (examples)

Chemical	Strength	Storage Concern	Restriction or Action
Caustic soda	50%	Freezing temperature is 54°F (12°C).	Dilute (caution heat) on receipt to drop freezing temperature. Prevent carbon dioxide absorption from air.
Sodium hypochlorite	10–20%	Degrades on lengthy storage, may form chlorate.	Monitor strength and limit storage time (60 days) or dilute on receipt. Monitor for chlorite and chlorate.
Liquid alum	40–50%	Diluted solution may hydrolyze.	Do not store diluted solution with pH more than 3.3.
Aqua ammonia	10–30%	Strength may change.	Keep cool and monitor strength.
Sodium chlorite	30–40%	Formation of chlorate and dry residue may be explosive.	Limit storage time, keep cool, use care to avoid dry residue formation.

Ops Tip

Alum used for jar testing should be diluted on day of use because pH may be greater than 3.3.

develop undesirable by-products when stored, and these must be monitored. Table 7-2 contains some common liquid water treatment chemicals with storage restrictions or special requirements. Consult the supplier's recommendations for proper storage of liquid treatment chemicals.

Liquid Chemical Feed Equipment

Chemical metering pumps are used to feed most liquid chemicals. Eductors and gravity feed systems are used in some cases, but a high percentage of the systems use a metering pump. Positive displacement pumps are used for this purpose because they deliver a constant flow to their maximum rated pressure (Figure 7-2).

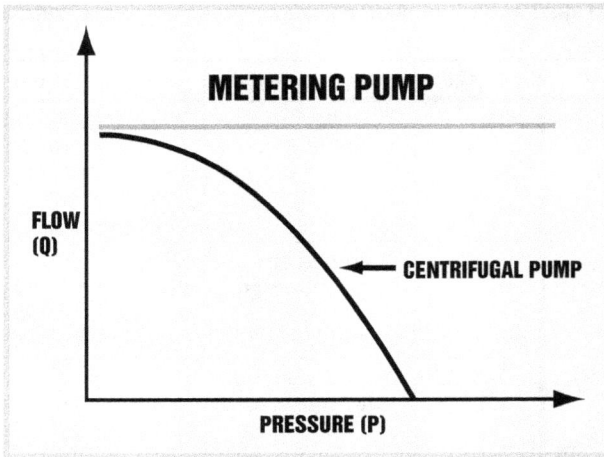

Figure 7-2   Metering pump flow compared to centrifugal pump

Several types of positive displacement pumps are used for chemi-cal feed. The more common types are
- Piston (Figure 7-3)
- Diaphragm (Figure 7-4)
- Gear (Figure 7-5)
- Peristaltic (tubing pumps)
- Syringe

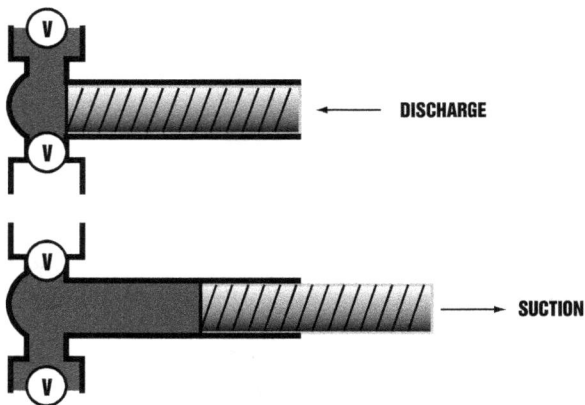

Copyright © by Cole-Parmer Instrument Company; used with permission.

Figure 7-3   Piston pump

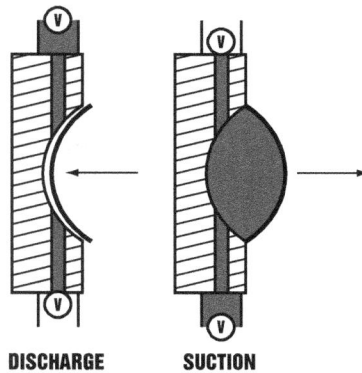

DISCHARGE   **SUCTION**

Copyright © by Cole-Parmer Instrument Company; used with permission.
Figure 7-4   **Diaphragm pump**

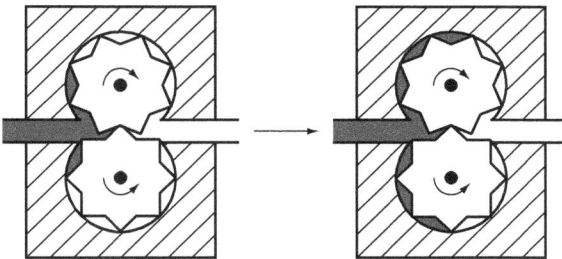

Copyright © by Cole-Parmer Instrument Company; used with permission.
Figure 7-5   **Gear pump**

Diaphragm pumps are used for a variety of chemical applications (Figure 7-4). Gear and piston pumps are sometimes preferred for thick liquids such as polymers. Peristaltic pumps are typically preferred for lower-pressure applications and are able to feed chemicals over a wide flow range because the treatment chemical contacts only a piece of tubing or a hose. Syringe pumps are suited to very low flows.

Diaphragm, piston, and peristaltic pumps all deliver liquids in pulses. This uneven delivery can cause dosing problems when mixing is not sufficient to ensure a rapid distribution of the chemical into the process water stream. Devices and systems are available to help smooth the delivery from these types of pumps. These devices are sometimes called *pulsation dampeners* (Figure 7-6). There are several designs.

One example is the gas bladder dampener. During the discharge stroke of a pump, fluid pressure takes the path of least resistance, displacing the bladder in the dampener, thus compressing the trapped gas. As the pump begins its next cycle, fluid flow stops momentarily, allowing the compressed gas to expand, forcing the bladder to push the accumulated fluid back into the discharge line, and filling the void created by the pump's cycle shift. This system evens out the flow from the pump.

The surfaces that contact the chemicals within the metering pump (wetted parts) must be chemically compatible. Aggressive treatment chemicals can require highly resistant materials, increasing the cost of the pump. Selecting the proper material of construction is critical to the performance and reliability of the pump. Follow manufacturer's recommendations and consult the chemical compatibility charts for guidance on correct materials for specific chemicals. Care should be taken when using existing chemical metering pumps, designed to feed one chemical, for another application. The new application may require another construction material. An incorrect choice of material can lead to pump failure.

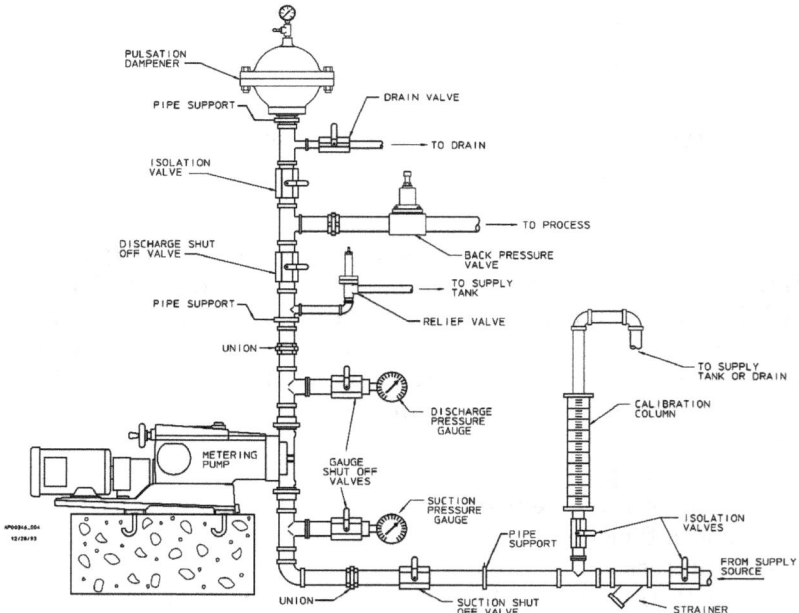

Courtesy of Pulsafeeder Inc.

Figure 7-6   Chemical metering pump installation with pulsation dampener

ALTERNATIVE 1

ALTERNATIVE 2

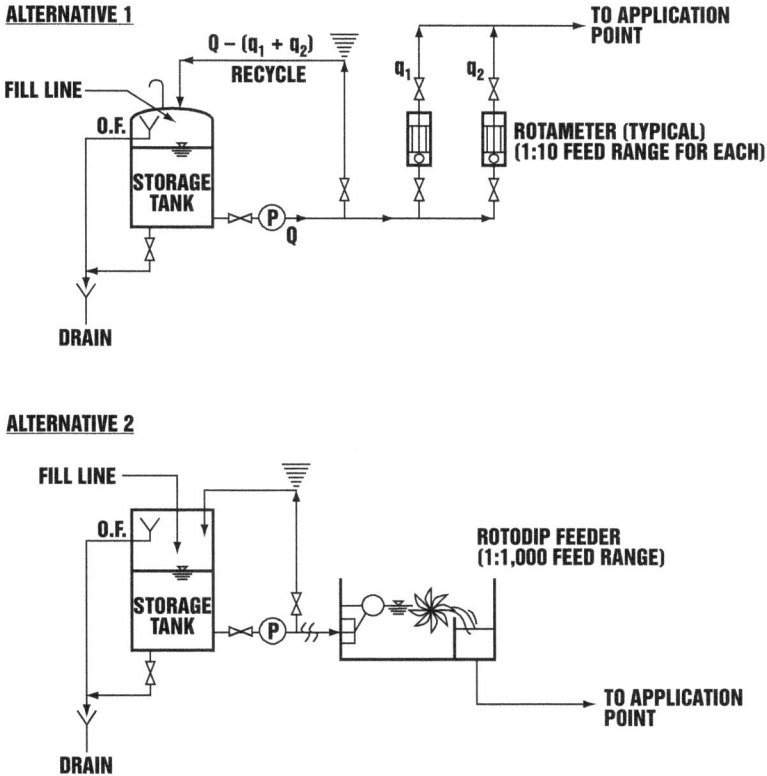

From Kawamura, Susumu, Integrated Design and Operation of Water Treatment Facilities, 2nd ed. Copyright © 2000 John Wiley & Sons, Inc. Reprinted with permission of John Wiley & Sons, Inc.

Figure 7-7   Liquid chemical feed systems

Liquid Chemical Feed Systems

Systems for feeding liquid chemicals consist of storage tanks (hazardous chemicals may require containment and neutralization facilities), metering pumps (with backflow prevention), and control instrumentation. Many possible system configurations exist. The feed application dictates the best choice for each installation. Two common systems are shown in Figure 7-7.

In some cases, a chemical day tank or dilution tank is provided (Figure 7-8). Utilities using large bulk storage tanks often use day tanks to minimize the risk of accidentally siphoning excessive quantities of treatment chemical because of a pump failure. Dilution tanks are needed when the chemical metering pump cannot deliver small quantities of the chemical for low-flow conditions. The chemical must then be diluted before it is fed through the metering pump. Another condition in which a day tank may be needed is where the bulk storage is a distance from the chemical pump location. In this case, periodically transferring undiluted chemical to the day tank may be the operating procedure. The day tank provides a reservoir of chemical near the pump, shortening the suction line. In addition, day tanks limit the amount of chemical lost to spillage when pump leaks or pipeline failures go unnoticed.

Solid (or dry) chemicals are sometimes used to prepare solutions for feeding. Many operators prefer to feed chemicals with metering pumps rather than dry feed equipment. Also, mixing liquid chemicals into the process water stream usually is more efficient. A weighted amount (using a continuous dry chemical feeder or a batch preparation approach) of the dry chemical is mixed with water to prepare the desired solution strength. Then the liquid is metered into the process water in the usual way.

Feed systems can be hybrids, combining dry or gas feed with liquid feed components. Dry and liquid systems usually include a dry-to-liquid conversion (using a dissolving tank) that subsequently combines this liquid with another liquid. Gas–liquid hybrid systems can mix the gas with water first or with another liquid chemical in a reaction chamber. Chlorine dioxide generators using chlorine gas are an example of this type of system. The chlorine dioxide formed in the reaction is ultimately mixed with ejector water, and this solution is then mixed with process water at the point of delivery.

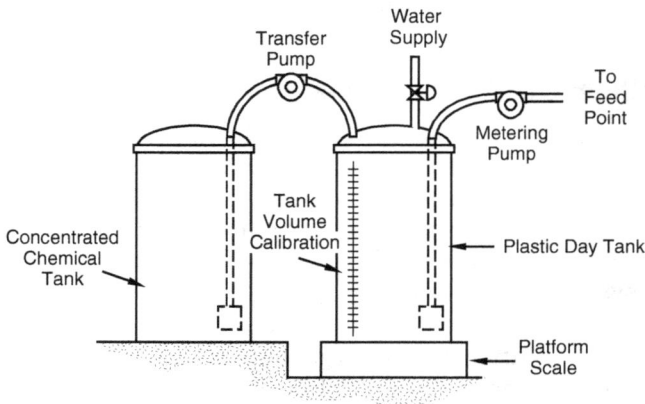

Figure 7-8   Chemical dilution tank

Ops Tip

Neutralize chlorite solution spills with sodium sulfite. *Do not use sodium bisulfite.*

Chlorine Dioxide Systems Using Liquid Generator Chemicals

On-site generation is necessary because the compressed liquid is explosive at room temperature. It is an unstable gas and therefore is not stored or shipped in bulk. Chlorine dioxide is most commonly produced by reacting chlorine gas with sodium chlorite. These generation systems are described in chapter 3, and they use various combinations of sodium chlorite, sodium chlorate, hydrogen peroxide, sulfuric acid, hydrochloric acid, and sodium hypochlorite solution.

Regulations limit the applied dosage of chlorine dioxide (the maximum contaminant level [MCL] is 0.8 mg/L) and the residual chlorite (1.0 mg/L MCL). Therefore, the dosage must be measured and controlled. It is important to measure the generator output chlorine

Courtesy David Tuck, Greenwood Commissioners of Public Works (S.C.).
Manufacturer: Severn Trent Services, ClorTec® Product Line.
Figure 7-9   On-site hypochlorite generation system

dioxide concentration. System calibration requires several generator settings to ensure accurate dosing. Measuring chlorine dioxide concentration after a predetermined contact time establishes the residual. This value verifies CT compliance (if it is used for primary disinfection) or is the control for the applied dosage.

Some chlorine dioxide feed systems use a batch generation process when filling a storage tank. A feed pump then meters the chemical to the feed location. Continuous chemical analyzers measure the chlorine dioxide concentration from the generator and the batch tank so the correct dosage matches the water flow. Also, chlorite (and chlorate) is monitored at several locations to ensure compliance with regulatory requirements.

On-Site Hypochlorite Generation and Feed Systems

The selection of liquid sodium hypochlorite rather than chlorine gas is often based on actual or potential costs to address safety issues. The Uniform Fire Code and OSHA requirements often increase the cost of using chlorine gas. Depending on a number of factors, gas containment and neutralization scrubbers may be needed. Also, there may be limits to storage inventory that require more frequent deliveries. Extensive emergency response plans may include elaborate evacuation plans and alert systems.

On-site hypochlorite generation systems (Figures 7-9 and 7-10) generally include water softener, a brine (salt water source) tank, electrolytic cells, and a hypochlorite storage tank. Sodium chloride (salt) is converted to hypochlorite by electrolysis. Some systems use seawater as the salt source, but it is common to make a brine solution

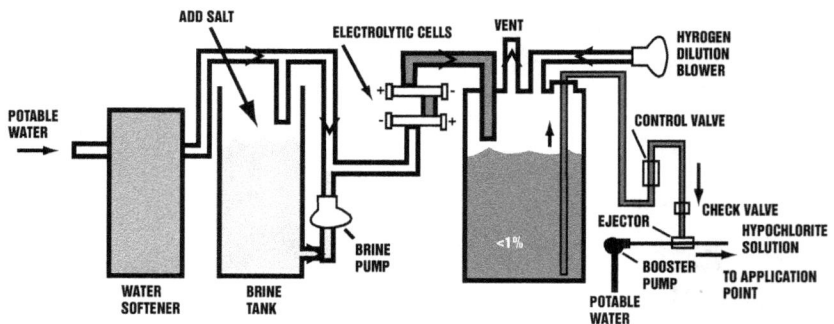

Used by permission of www.globaltreat.com

Figure 7-10   On-site hypochlorite generator system

using commercial salt. The hypochlorite solution is usually fed to the point of application by a chemical feed pump, similar to other liquid chemicals.

Most hypochlorite generation systems produce low-concentration solutions (around 1 percent hypochlorite). Some systems produce higher concentrations and some membrane systems even generate chlorine gas and sodium hydroxide (chlor–alkali cells). Several proprietary systems produce either chlorine or a liquid solution of mixed oxidants (e.g., ozone, chlorine dioxide, and chlorine). Most of these systems employ salt brine and an electrolytic cell to produce the chemical product(s).

Polymer Feed Systems

Polyelectrolytes (or polymers) used in water treatment systems have specific storage, handling, feeding, and dilution requirements. These chemicals are typically fed in low concentration, so it is imperative that these materials be fed accurately to prevent underfeeding and overfeeding, which can result in wasted chemical treatment and poor system performance.

Polymer Types

Polymers are available as powders, liquids, and emulsions. Each form has different feeding, handling, and storage requirements.

Both cationic and anionic high-molecular-weight polymers are available in powdered form. These products have the advantage of being 100 percent polymer, which can minimize shipping and handling costs. However, it is absolutely essential that dry polymer materials be handled and diluted properly to prevent underfeeding and overfeeding.

Solution polymers are usually cationic, low-molecular-weight, high-charge-density products, and are commonly used for coagulation or coagulation aids. Solution polymers are easier to dilute, handle, and feed than dry and emulsion polymers. In many cases, predilution of a solution polymer is unnecessary, and the product can be fed directly from the shipping container or bulk storage tank. They also can be diluted to any convenient strength consistent with chemical feed pump output.

High-molecular-weight polymers are also available as emulsions. An emulsion product allows the manufacturer to provide concentrated liquid polymer formulations that cannot be made in solution form. It is only after the emulsion polymer is activated with water that the polymer is available in its active form. Therefore, these products must be diluted properly prior to use.

Storage

Dry polymers are susceptible to caking if stored in highly humid conditions. Caking is undesirable because it interferes with the dilution process. Therefore, dry polymers should be kept in areas of low humidity, and opened containers of dry material should be resealed. Most polymer products begin to lose their activity after one year of storage. Although this process is gradual, it ultimately affects the cost of chemical treatment. Therefore, polymers should be used before their expiration date.

Solution polymers should be stored in an area protected from freezing. Some solution products are susceptible to irreversible damage when frozen. Others exhibit excellent freeze–thaw recovery. In no case should solution polymers be stored at temperatures above 120°F. As solutions, these polymers do not require periodic mixing (to prevent separation) before use. However, some solution polymers have a short shelf life, and inventory should be adjusted accordingly.

Because emulsion polymers are not true solutions, they separate if allowed to stand for a prolonged period of time. Therefore, emulsion polymers must be mixed with a drum mixer, tank mixer, or tank recirculation package prior to use. A bulk tank or bin recirculation package should be designed to circulate the tank's contents at least once per day to prevent separation. Emulsion polymers contained in drums should also be mixed daily.

Neat emulsion polymer must be protected from water contamination, which causes gelling of the product and can make pumping difficult or impossible. In areas of high humidity, tank vents should be outfitted with a desiccant in order to prevent water condensation within the emulsion storage tank. Even small amounts of condensation can cause significant amounts of product gelling. As with liquid

Ops Tip
Excessive mixing may damage the polymer, so be sure to consult with the manufacturer for guidance.

products, emulsion polymers must be protected from freezing and should be stored at temperatures below 120°F.

Dilution and Feeding

Dry polymers must be diluted with water before use. Most operations require preparation of polymer dilutions once per shift or daily. Typically, a plant operator is charged with the responsibility of measuring a correct amount of dry polymer into a container. The contents of the container are conveyed to the mixing tank through a polymer eductor.

If dry polymer particles are not wetted individually before introduction into the dilution tank, "fisheyes" (undissolved globules of polymer) will form in the solution tank. Fisheyes represent wasted polymer and cause plugging in chemical feed pumps.

Dry polymer solution strengths must be limited to approximately 0.5–1 percent or less by weight, depending on the product used. This is necessary to keep the solution viscosity (thickness) to a manageable level. The mixer employed in the solution tank should not exceed 350 rpm, and mixing should last only until all material is dissolved. Normally, a batch of diluted dry polymer should be used within 24 hr of preparation, because the diluted product begins to lose activity after this amount of time.

Automatic dry polymer dilution systems can be used to perform the wetting, diluting, and mixing functions; however, the system must be manually recharged with dry polymer periodically. Although costly, these systems can save time and usually improve consistency.

Solution polymers may be diluted prior to use or fed neat from a shipping container, bin, or bulk storage tank. Dilution of these products becomes necessary if there is insufficient mixing available to combine the polymer with the water being treated. In-line static mixer dilution systems are acceptable for solution polymers and are the simplest method of solution polymer dilution and feed. A solution polymer can be fed through one of the many commercially available emulsion polymer dilution systems.

Solution polymers can be pumped most easily with gear pumps. However, many solution polymers have a viscosity low enough to be pumped by diaphragm chemical metering pumps.

Emulsion polymers must be diluted before use. Dilution allows the emulsion product to invert and "converts" the polymer to its active state. Proper inversion of emulsion polymers is rapid and effective. Improper inversion of the emulsion polymer can result in loss of activity caused by incomplete uncoiling and dissolution of the polymer molecules.

Batch and continuous make-down systems are acceptable for emulsion polymer use. In batch preparation, a plant operator feeds a premeasured amount of neat emulsion product into the agitator vortex of a dilution tank. The product is mixed until it is homogeneous, and then the mixers are shut off. As with dry polymer products, mixer speed should always be below 350 rpm, and the mixer should be shut off as soon as the product is homogeneous. This prevents excessive shearing of the polymer molecule and resultant loss of polymer activity.

Several manufacturers market continuous emulsion polymer feed systems. These systems pump neat polymer from the storage container into a dilution chamber, where the polymer is combined with water and fully activated. The polymer–water solution then flows by water pressure to the point of application. Provision is made for secondary in-line dilution water to dilute the polymer further prior to use. These polymer feed systems are by far the easiest and best ways to feed emulsions continuously.

It is not acceptable to use in-line static mixing alone for dilution of emulsion polymers. However, in-line static mixing can be employed for blending secondary dilution water with diluted emulsion product prior to application. Initial dilution of emulsion polymers should be 1 percent or 2 percent by weight. This solution strength ensures proper particle-to-particle interaction during the inversion step, which aids in complete inversion.

It is usually desirable to provide secondary dilution water capabilities to emulsion polymer feed systems, because these products tend to be most effective when fed at approximately 0.1 percent solution strength.

General Recommendations

Additionally, some general guidelines apply to the feeding and handling of all water treatment polymers. In areas where the temperature routinely drops below freezing, it is good practice to insulate all polymer feed lines.

For tank batches of diluted polymers, tank mixer speeds of more than 350 rpm should not be used. In the preparation of diluted batches of polymer, water should always be added to the tank first. Then, the mixer should be started and the polymer added on top of the water.

Diaphragm metering pumps can be used to pump most polymer

solutions. However, because of the viscosity of some products, gear pumps may be necessary. Plastic piping should be used in polymer feed systems; stainless steel is also acceptable. Most polymers are corrosive to mild steel and brass. Extra precautions should be taken to prevent spilling of polymers, because wet polymer spills can become extremely slippery and present a safety hazard. Spills should be covered with absorbent material, and the mixture should be removed promptly and disposed of properly.

Metering Pump Calibration

Periodically calibrating chemical metering pumps is necessary to ensure accurate chemical feed. The frequency of calibration depends on the stability of the pump, the chemical being fed, and the impact of changing treatment conditions. Experience with a chemical feed system will support the need for a specific calibration routine. Most chemical metering pumps should be calibrated at least monthly. Even digital precalibrated chemical metering pumps should have the calibration validated from time to time.

A chemical metering pump should have a calibration chamber installed (Figures 7-11, 7-12, and 7-13). This facilitates periodic calibration. Pump calibration can be performed without a permanently installed calibration system. This would require valves to discharge the chemical from the pump to a container. The volume delivered for a measured time can then be recorded.

Many metering pumps have adjustments for both stroke and speed. The stroke is the amount of chemical pumped for each pump cycle. The speed adjusts the frequency for each stroke (i.e., strokes per minute). Most control systems can automatically, or at least remotely, adjust the pump speed, and the stroke is set manually. This is the reason that most calibration curves show the volume delivered for various pump speeds with the stroke fixed for a few specific settings (Figure 7-4 shows an example of an alum pump calibration).

The calibration procedure for a chemical metering pump involves several steps.

1. Set the pump speed and stroke to begin the procedure.

2. Open valve(s) to fill calibration chamber with chemical.

Ops Tip

Even digital pumps must be calibrated.

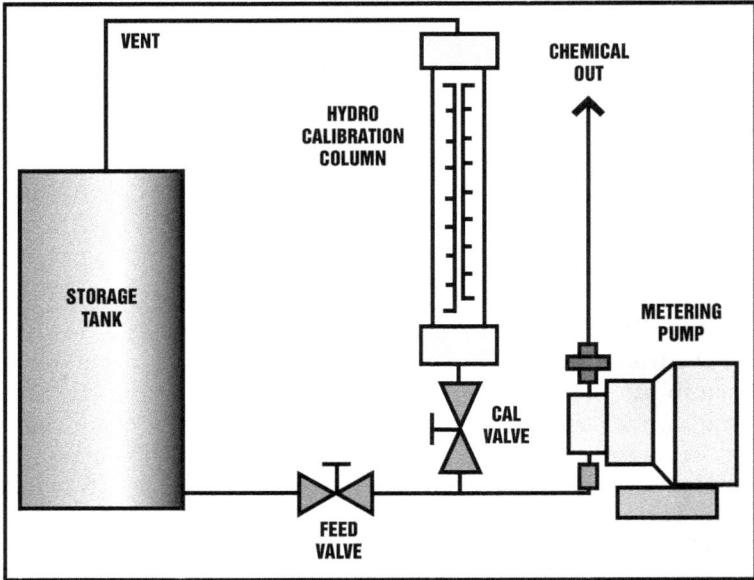

Courtesy of Hydro Instruments

Figure 7-11   Metering pump calibration chamber installation

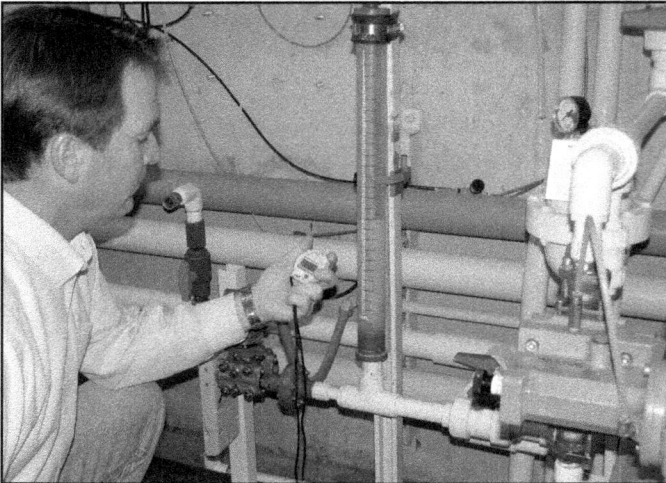

Courtesy of David K. Hardy

Figure 7-12   Calibrating a metering pump

Courtesy of Bill Soucie, CLCJAWA
Figure 7-13   Metering pump with calibration chamber

3. Fill calibration chamber to a point above a major measurement value.

4. Isolate metering pump to pump from calibration chamber.

5. Zero timer. Begin pumping from the calibration chamber. When chemical drops to gradation, begin timer.

6. Pump chemical for a measured time period; the longer, the more accurate.

7. Note volume of chemical pumped and time; record result.

8. Select different pump settings and record values; repeat calibration procedure.

9. Plot results on chart to develop calibration curves (Figure 7-14).

Constructing a Calibration Chart

A metering pump calibration chart is constructed by plotting the results from the calibration procedure previously described. The pump rate (mL/min) should be plotted for each pump speed setting, and a line should be drawn through the points on graph paper or using a charting program (e.g., Microsoft Excel). First, the pump rate (mL/min) should be calculated for each timed volume measured for the calibration of the pump. If the volume is collected (pumped) for a time other than 1 minute, these values will need to be calculated (c7-1). Table 7-3 lists pump rate values for volumes pumped in 10 seconds.

c7-1   60 × | Volume pumped (mL) | ÷ | Time (sec) | = | Pump rate (mL/min) |

Table 7-3   Pump rate conversion

Volume (mL) Pumped in 10 sec	Pump Rate (mL/min)
10	60
20	120
30	180
40	240
50	300
60	360
70	420
80	480
90	540
100	600
250	1,500
500	3,000
1,000	6,000
1,500	9,000
2,000	12,000
3,000	18,000

Example 7-1   An alum feed pump is calibrated as follows:

Calibration Results

Pump Speed Setting	Pump Stroke Setting	Volume Pumped (mL)	Time (sec)	Pump Rate (mL/min)
25	50	104	10	624
50	50	200	10	1,200
75	50	317	10	1,902
100	50	400	10	2,400

Plot these results on a chart. See Figure 7-14 for an example. Repeat this procedure for other speed and stroke settings.

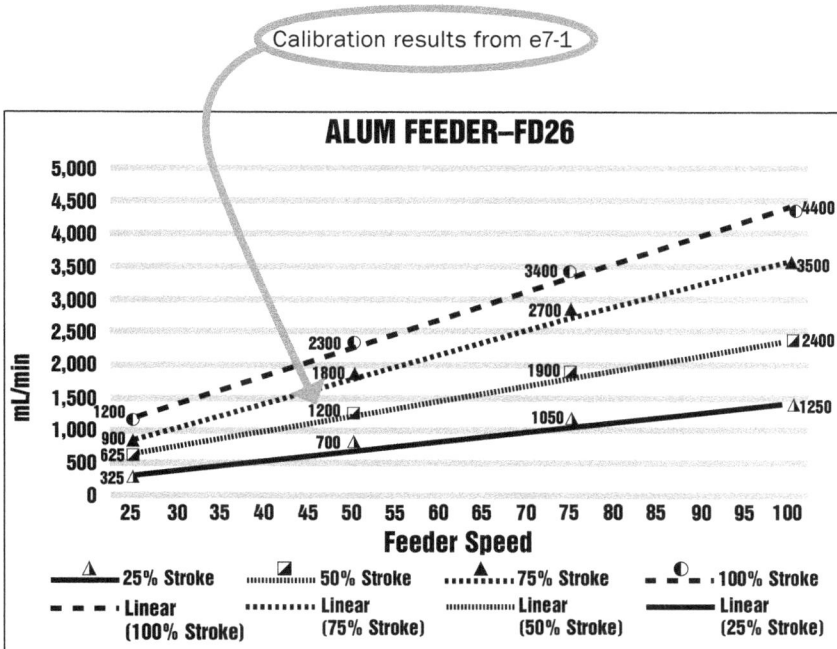

Calibration results from e7-1

ALUM FEEDER–FD26

mL/min		
5,000		
4,500		4400
4,000		
3,500	3400	3500
3,000	2700	
2,500		2400
2,000	2300 1800	1900
1,500 1200	1200	1250
1,000 900	700	1050
500 625		
325		
0		

Feeder Speed: 25 30 35 40 45 50 55 60 65 70 75 80 85 90 95 100

25% Stroke   50% Stroke   75% Stroke   100% Stroke
Linear (100% Stroke)   Linear (75% Stroke)   Linear (50% Stroke)   Linear (25% Stroke)

Figure 7-14   Alum pump calibration chart

Feeding Liquid Chemicals in Water Treatment

Many chemicals used in water treatment are liquids. Operators often consider liquids for safety and ease of operation. The chemical feed systems vary in complexity. Simple systems feed the liquid directly to the process water stream without any dilution. Complex systems can combine gases or solids with liquids to form treatment chemicals. Batches of various strengths may be used to improve chemical feed accuracy and aid in dispersion of the material into a variable flow of water. Most water treatment plants feed at least some liquid chemicals as part of their treatment process.

Chapter 8

Liquid Chemical Feed— Calculations and Examples

Many chemicals used in water treatment are liquids (Table 1-1). Some can be fed directly into the process water stream where they are vigorously mixed. Many times, highly concentrated liquid chemicals are first diluted before metering the chemical to the point of delivery. Dilution is accomplished on delivery for some chemicals and in batches for others. Although chemicals can be fed using eductors or even gravity, it is more common and more accurate to use metering pumps for this purpose. It is critical to periodically calibrate metering pumps to ensure accurate chemical feed.

Direct Feed Calculations

Some liquid chemicals are fed directly from the storage tank to the process water without any dilution. The chemicals may be delivered already diluted or may be 100 percent strength. Calculating the chemical feed rate requires knowing the process water flow rate, the applied dosage, and the solution strength.

Calculators 8-1 and 8-2 (identical to dry feed calculators 6-1 and 6-2) can be used to determine the feed rate of "pure" chemical for a given dosage and flow rate. These are often used for liquid polymers (polyelectrolytes) where the polymer in the delivery drum is fed directly to the point of application without dilution.

c8-1
$$\boxed{\text{Feed rate (lb/day)}} = 8.34 \times \boxed{\text{Dosage (mg/L)}} \times \boxed{\text{Water flow (mgd)}}$$

c8-2
$$\boxed{\text{Feed rate (g/hr)}} = 41.7 \times \boxed{\text{Dosage (mg/L)}} \times \boxed{\text{Water flow (ML/day)}}$$

However, even in this case, the feed rate may need to be in volume rather than weight. Calculators c8-3 and c8-4 can be used for this calculation (Table 8-1). The density (weight/volume) or specific gravity must be known for this calculation. The manufacturer should supply this value. A good approximation of this value can be acquired by weighing a known volume of the material. Calculator c8-5 should be used for the calculation.

c8-3 $\boxed{\text{Feed rate (gpd)}} = 0.12 \times \boxed{\text{Feed rate (lb/day)}} \div \boxed{\text{Specific gravity}}$

c8-4 $\boxed{\text{Feed rate (L/day)}} = 0.024 \times \boxed{\text{Feed rate (g/hr)}} \div \boxed{\text{Specific gravity}}$

c8-5	[chemical weight (g) of 100 mL of chemical] ÷ 100 = specific gravity (estimate)
	Calculator instructions: A cylinder and a scale are needed. Weigh a graduated cylinder, add 100 mL of chemical (exactly), weigh again (the bottom of the meniscus should rest on the top of the 100 mL gradation), and subtract weight of cylinder from total; this is the weight of 100 mL of chemical. The weight of 100 mL of water is 100 g. The specific gravity is the weight of the chemical divided by the weight of water.

Table 8-1   Conversion factors for common feed rate units

Convert From	Multiply by	Converted to
gpd	.0007	gal/min
gpd	2.63	mL/min
lb/d	.042	lb/hr
lb/hr	.017	lb/min
lb/d	.0007	lb/min
g/hr	.017	g/min
g/hr	.024	kg/d
L/d	0.69	mL/min
L/d	.042	L/hr
L/d	0.001	m^3/d

**Table
Tamer**   To convert in the reverse direction, divide by the conversion factor.

Example: Convert 52 gpd to mL/min
Multiply 52 gpd by 2.63 = **136.8 mL/min**
Convert 2.5 g/min to g/hr
Divide 2.5 g/min by 0.017 = **147.1 g/hr**

Diluted Chemical Feed Calculations

Liquid chemicals are often diluted prior to being fed to the process water stream. Diluted chemicals are used for several reasons.

1. Diluted chemicals mix better with process water.

2. Chemical feed pumps may operate in the optimum range (rather than at the extreme low end of adjustment).

3. Viscosity of the concentrated chemical may present pumping problems (some polymers have this property).

4. Concentrated chemical may freeze or crystallize at relatively normal temperatures.

5. Corrosive properties of the concentrated chemical may require storage in special containers.

6. Bulk chemical storage tanks may be distant from chemical feed pump location. Diluted chemical day tanks may provide local supply for continuous pumping.

7. Chemical feed solutions are prepared from dry or gaseous pure chemicals. Therefore, these solutions are less than full strength.

8. Some chemical solutions are prepared by on-site generation equipment. This results in a diluted solution for chemical feed to the process water.

In these situations, the chemical is not full strength, and this must be considered in the dosage calculations.

When feeding diluted chemical solutions, two primary values that most operators need to calculate are the chemical pump feed rate and the chemical usage (for inventory control). Calculators c8-6 through c8-8 should be used to determine these results.

Preparing Diluted Chemical Solutions

Solution Strength Percentages
- %w/w = weight of chemical to total weight of solution
- %w/v = weight of chemical to volume of solution, called *trade percent*
- %v/v = volume of chemical to volume of solution

Calculators in this field guide use w/w for solution strength percentage.

Solution strength is usually expressed as a percent by weight. This is the weight of dry (or pure) chemical used for each pound (or gram) of water to make the solution. This can be confusing, because solutions often are measured in volume rather than weight. So this requires some conversion (volume to weight and back again).

To illustrate, a 15 percent alum solution should be prepared using pure aluminum sulfate (dry alum) and water. A gallon of water weighs 8.34 lb, so for each gallon (8.34 lb), 15 percent of that amount of dry alum (1.25 lb) must be used. The calculators for the amount of dry chemical needed to prepare any volume of any strength solution are therefore (c8-6 and c8-7) as follows.

c8-6 $\boxed{\text{Dry chemical (lb)}} = 0.0834 \times \boxed{\text{Solution strength (\%)}} \times \boxed{\text{Volume (gal)}}$

c8-7 $\boxed{\text{Dry chemical (kg)}} = 0.01 \times \boxed{\text{Solution strength (\%)}} \times \boxed{\text{Volume (L)}}$

In some cases, the full-strength chemical is a liquid instead of a dry chemical. Some polymers fall into this category. It is convenient to measure these by volume, so it is necessary to know the specific gravity or density to make a conversion from volume to weight (c8-8 and 8-9). Many other chemicals are delivered and stored as concentrated solutions. These are then diluted before feeding the chemical. The following calculators can be used to calculate the volume of the concentrated chemical to make a given volume of a certain strength. Again, the density or specific gravity of the concentrated (original) chemical must be known.

c8-8 $\boxed{\text{Liquid chemical volume (gal)}} = 0.0834 \times \boxed{\text{Solution strength (\%)}} \times \boxed{\text{Volume (gal)}} \div \boxed{\text{Density of chemical (lb/gal)}}$

c8-9 $\boxed{\text{Liquid chemical volume (L)}} = 10 \times \boxed{\text{Solution strength (\%)}} \times \boxed{\text{Volume (L)}} \div \boxed{\text{Density of chemical (g/L)}}$

Example 8-1 Calculate the volume of liquid polymer needed to make 500 gal of a 1 percent solution used for chemical feed. The density of the original liquid polymer is 9.5 lb/gal.

Using calculator c8-8   **1** ▸   **500** ▸   **9.5** ▸

$\boxed{\text{Liquid chemical volume (gal)}} = 0.0834 \times \boxed{\text{Solution strength (\%)}} \times \boxed{\text{Volume (gal)}} \div \boxed{\text{Density of chemical (lb/gal)}}$

Volume of liquid polymer (gal)   = $0.0834 \times 1\% \times 500$ gal $\div$ 9.5 lb/gal

        = 4.39 gal of liquid polymer

Relationship of Density to Specific Gravity

Density of a substance is the weight for a unit volume. For liquids used in water treatment, this is expressed as pounds per gallon (lb/gal) or grams per milliliter (g/mL or g/cm^3). Specific gravity is the relative density of a liquid (in this case) compared to the density of water at the same temperature. Specific gravity, therefore, does not have any units associated with it. Substances that are denser than water have a value greater than one.

Because the solutions used in water treatment are mostly water, the density of the substance can be estimated from the specific gravity (with only a small error in most cases). The density of water is 8.34 lb/gal or 1 g/mL.

1. Specific gravity = Density of solution/Density of water, or

2. Density of solution = Specific gravity × Density of water

Expressing density can lead to some confusion. The reason for this confusion is the units lb/gal or g/mL are also used to describe the weight of dry chemical in the solution.

Liquid alum can be used as an example. Most liquid alum is delivered as 48 percent strength (specific gravity of 1.33). The density of this solution is therefore 1.33×8.34 lb/gal = **11.1 lb/gal**. This is the weight of 1 gal of 48 percent strength liquid alum.

> **Example 8-2**   Calculate the density of an alum solution where the specific gravity is 1.3063.
>
> Density of solution = Specific gravity × Density of water
>
> Density of solution = 1.3063 × 8.34 = **10.89 lb/gal**

The weight of dry alum used to make 1 gal of this solution is then 11.1 lb/gal × 0.48 = **5.33 lb/gal.** This value is used in the chemical feed calculation because it is the amount of dry (pure) chemical that is used to determine the dosage (if fed as dry alum). The chemical feed calculators use either the solution density or the specific gravity, and the solution strength. Most operators use the specific gravity and the solution strength for their calculations. Solution density is sometimes known (and specific gravity is not); this and the solution strength are then used in the calculation.

Chemical Feed of Solutions

Determining the chemical feed rates for diluted chemical solutions (c8-7 and c8-8) requires knowing the water flow rate, the strength of the diluted chemical solution, and the chemical feed dosage. Most liquid chemical metering pump feed rates are expressed in mL/min. For inventory control or chemical usage over a protracted period, gallons per day (gpd), or liters per day (L/d), may be a more useful unit of measure.

	Chemical feed rate (mL/min)		Chemical dosage (mg/L)	Water flow (mgd)	Solution strength (%)	Density of chemical (lb/gal)
c8-10		= 2,192 ×	×	÷	÷	

	Chemical feed rate (mL/min)		Chemical dosage (mg/L)	Water flow (ML/day)	Solution strength (%)	Density of chemical (g/ML)
c8-11		= 69.4 ×	×	÷	÷	

The *density of chemical* is the weight/volume of the chemical being fed. This can be determined from manufacturer's information or by calculation from measuring the specific gravity (Example 8-2). The solution strength in percent is obtained from manufacturer's information (or from provided specific gravity tables for each chemical; see Appendix C for examples).

Chemical feed rate calculators c8-12 and c8-13 use specific gravity instead of the density. These are more useful to many operators because they often measure the specific gravity for quality control purposes. The metric calculator (c8-13) is the same as c8-8 because the density of the chemical in g/mL is the same as the specific gravity for most water solutions.

c8-12 $\boxed{\begin{array}{c}\text{Chemical}\\\text{feed rate}\\\text{(mL/min)}\end{array}} = 262.9 \times \boxed{\begin{array}{c}\text{Chemical}\\\text{dosage}\\\text{(mg/L)}\end{array}} \times \boxed{\begin{array}{c}\text{Water}\\\text{flow}\\\text{(mgd)}\end{array}} \div \boxed{\begin{array}{c}\text{Solution}\\\text{strength}\\\text{(\%)}\end{array}} \div \boxed{\begin{array}{c}\text{Specific}\\\text{gravity of}\\\text{chemical}\\\text{solution}\end{array}}$

c8-13 $\boxed{\begin{array}{c}\text{Chemical}\\\text{feed rate}\\\text{(mL/min)}\end{array}} = 69.4 \times \boxed{\begin{array}{c}\text{Chemical}\\\text{dosage}\\\text{(mg/L)}\end{array}} \times \boxed{\begin{array}{c}\text{Water}\\\text{flow}\\\text{(ML/day)}\end{array}} \div \boxed{\begin{array}{c}\text{Solution}\\\text{strength}\\\text{(\%)}\end{array}} \div \boxed{\begin{array}{c}\text{Specific}\\\text{gravity of}\\\text{chemical}\\\text{solution}\end{array}}$

Chemical Feed Calculations for Very Dilute Solutions: Special Case

Chemical feed rate calculators c8-14 through c8-17 are for use with very dilute solutions (<2 percent strength by weight). For these calculations, the density (specific gravity) is estimated to be the same as water (this causes only a small error). The concentration of the solution may be obtained from test results, and then the strength of the solution is given in mg/L rather than percent. These calculators are often used for on-site hypochlorite generators (concentration is approximately 1 percent), chlorine dioxide generators (where the generator output concentration is measured in mg/L), or dilute polymer feed solutions.

c8-14 $\boxed{\begin{array}{c}\text{Chemical}\\\text{feed rate}\\\text{(mL/min)}\end{array}} = 262.9 \times \boxed{\begin{array}{c}\text{Chemical}\\\text{dosage}\\\text{(mg/L)}\end{array}} \times \boxed{\begin{array}{c}\text{Water flow}\\\text{(mgd)}\end{array}} \div \boxed{\begin{array}{c}\text{Solution}\\\text{strength}\\\text{(\%)}\end{array}}$

c8-15 $\boxed{\begin{array}{c}\text{Chemical}\\\text{feed rate}\\\text{(mL/min)}\end{array}} = 69.4 \times \boxed{\begin{array}{c}\text{Chemical}\\\text{dosage}\\\text{(mg/L)}\end{array}} \times \boxed{\begin{array}{c}\text{Water flow}\\\text{(ML/d)}\end{array}} \div \boxed{\begin{array}{c}\text{Solution}\\\text{strength (\%)}\end{array}}$

c8-16 $\boxed{\begin{array}{c}\text{Chemical}\\\text{feed rate}\\\text{(mL/min)}\end{array}} = 2,629,000 \times \boxed{\begin{array}{c}\text{Chemical}\\\text{dosage}\\\text{(mg/L)}\end{array}} \times \boxed{\begin{array}{c}\text{Water}\\\text{flow}\\\text{(mgd)}\end{array}} \div \boxed{\begin{array}{c}\text{Solution}\\\text{strength}\\\text{(mg/L)}\end{array}}$

c8-17 $\boxed{\begin{array}{c}\text{Chemical}\\\text{feed rate}\\\text{(mL/min)}\end{array}} = 694,000 \times \boxed{\begin{array}{c}\text{Chemical}\\\text{dosage}\\\text{(mg/L)}\end{array}} \times \boxed{\begin{array}{c}\text{Water flow}\\\text{(ML/d)}\end{array}} \div \boxed{\begin{array}{c}\text{Solution}\\\text{strength}\\\text{(mg/L)}\end{array}}$

Chemical Feed Rate Calculation Examples

Example 8-3   Calculate the feed rate for liquid alum in mL/min.

Plant flow =	25 mgd
Alum dosage =	17.0 mg/L
Alum solution strength =	48%
Alum solution specific gravity =	1.29

The values are in US units, so use calculator c8-12.

		17.0	25	48	1.29
Chemical feed rate (mL/min)	= 262.9 ×	Chemical dosage (mg/L)	× Water flow (mgd)	÷ Solution strength (%)	÷ Specific gravity of chemical solution

mL/min = 262.9 × 17.0 × 25 ÷ 48 ÷ 1.29

= 1,804 mL/min

Example 8-4   Calculate the liquid alum feed rate in mL/min.

Plant flow =	3 ML/d
Alum dosage =	15.0 mg/L
Alum solution strength =	47.8%
Alum solution specific gravity =	1.23 g/mL

The values are in metric, so use calculator c8-13.

		15.0	3	47.8	1.23
Chemical feed rate (mL/min)	= 69.4 ×	Chemical dosage (mg/L)	× Water flow (ML/d)	÷ Solution strength (%)	÷ Density of chemical (g/mL)

mL/min = 69.4 × 15.0 × 3 ÷ 47.8 ÷ 1.23

= 53.1 mL/min

Ops Tip
Make sure that all of the units match with the calculator before using the formula.

Calculation Starting with Different Units

To use the calculators, all of the units must match those in the calculator. If some of the units are different, they must be converted so that they match. The following is an example.

Example 8-5   Calculate the liquid alum feed rate in mL/min if

Plant flow =	205 gpm
Alum dosage =	1.2 grains/gal
Alum solution strength =	48%
Alum solution specific gravity =	1.29

Units are US and the specific gravity is provided, so use calculator c8-12.

		20.5		0.30		48		1.29
Chemical feed rate (mL/min)	= 262.9 ×	Chemical dosage (mg/L)	×	Water flow (ML/d)	÷	Solution strength (%)	÷	Specific gravity of chemical solution

1. The water flow is in gpm and must be converted to mgd. Multiply gpm by 0.00144 to convert to mgd.

   205 gpm × 0.00144 = 0.30 mgd

2. The chemical dosage is in grains/gal and must be converted to mg/L. Multiply grains/gal by 17.1 to convert to mg/L.

   1.2 grains/gal × 17.1 = 20.5 mg/L

Now all of the necessary information is in the correct units. Insert the previous values and complete the calculation.

mL/min   = 262.9 × 20.5 × 0.30 ÷ 48 ÷ 1.29

= 26.1 mL/min

Example 8-6   Calculate the liquid sodium hypochlorite feed rate in mL/min if

Water flow =	350 gpm
Chlorine dosage =	3.0 mg/L
Hypochlorite solution strength =	12% trade
Hypochlorite solution specific gravity = 1.174	

Units are US and the specific gravity is provided. Use calculator c8-12.

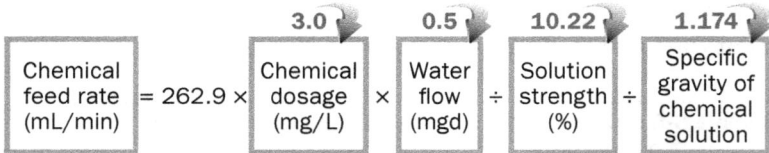

	3.0	**0.5**	**10.22**	**1.174**
Chemical feed rate (mL/min) = 262.9 ×	Chemical dosage (mg/L) ×	Water flow (mgd) ÷	Solution strength (%) ÷	Specific gravity of chemical solution

1. Convert gpm water flow to mgd to match calculator.

gpm × 0.00144 = mgd

350 gpm × 0.00144 = **0.5 mgd**

2. Convert trade percent to solution strength percent.

Consult hypochlorite specific gravity table provided by manufacturer (Appendix C, Table C-4).

12% trade = **10.22%** available chlorine

All units now match the calculator. Substitute the values into the calculator.

mL/min   = 262.9 × 3.0 × 0.5 ÷ 10.22 ÷ 1.174

= 32.9 mL/min

Ops Tip   Percent trade is g/100 mL (%w/v) and solution strength percent is g/100 g (%w/w). The %w/w can be found in the sodium hypochlorite table (C-4) in Appendix C or calculated by dividing the percent trade by the specific gravity (i.e., 12%/1.174 = 10.22% by weight.)

Inventory Management for Liquid Chemicals

Daily chemical usage trends are used to make reorder decisions. Carefully managing chemical inventory ensures adequate supply without requiring excess storage capacity. This reduces costs and helps maintain chemical strength and reduce unwanted by-products sometimes formed during prolonged storage. Liquid chemicals are particularly vulnerable to these product changes.

The chemical feed rate in mL/min is easily converted to gpd (kL/day) using conversion factors shown in Table 8-2.

Table 8-2   Chemical feed daily usage conversion factors

Convert from	Multiply by	Convert to
mL/min	1.44	L/day
mL/min	0.00144	m³/day
mL/min	0.38	gpd
mL/min	0.054	gal/wk

Example 8-7   Calculate the daily liquid alum usage if the chemical feed rate is 1,805 mL/min.

1,805 mL/min × 0.38 = **685.9 gpd** or

1,805 mL/min × 1.44 = **2,599 L/day** or

1,805 mL/min × 0.00144 = **2.6 m³/day**

SCADA systems are easily configured to plot chemical usage trends and to estimate chemical order dates. Liquid chemical usage can be tracked manually by plotting daily usage on a chart (Figure 8-1). A trendline can be drawn through the data points and extended into the future, and this can be used to estimate projected order dates. Safety storage amounts can be added and days included to account for delivery time, including holidays and weekends. Also, operators should consider the necessary storage to accept full truck loads (depending on the chemical, this can be 5,000 gal) to take advantage of any bulk order shipping discounts.

Table 8-3 shows the daily chemical usage, the running total chemical usage, and the usage deducted from initial amount in storage. The total chemical usage deducted from the amount in storage is plotted daily (Figure 8-1). A trendline is drawn through the daily usage values and extended to zero in storage. Then, the reorder date is estimated by backing off the days for emergency storage, days for delivery, and any other days needed. The amount in storage and the amount of storage space available on the estimated reorder date are then checked to make sure these are adequate.

Once the new load of chemical is received, it is added to the storage amount at that time and the process is started over. This is a useful method to track chemical usage and to help determine the best time to reorder chemicals.

Table 8-3   Liquid chemical usage and deduction from storage

Date	Usage (gal/d)	Accumulated Chemical Usage (gal)	Total Accumulated Usage Deducted from Storage (gal)
1	679	679	18,000
2	567	1,246	16,754
3	478	1,724	16,276
4	789	2,513	15,487
5	356	2,869	15,131
6	777	3,646	14,354
7	679	4,325	13,675
8	444	4,769	13,231
9	666	5,435	12,565
10	567	6,002	11,998
11	876	6,878	11,122
12	956	7,834	10,166
13	408	8,242	9,758

The beginning amount of chemical in storage on day 1 in Table 8-3 is 18,000 gallons. This plant had two 12,000-gal storage tanks, so one was full and one was half full. The rate that the chemical is used varies daily from 356 gal to 956 gal. However, when these amounts are deducted from the initial storage amount (18,000 gal) and plotted on a graph (Figure 8-1), the rate of usage is fairly uniform. The trendline is easy to draw through the points and beyond. The figure indicates that ordering a 5,000-gal delivery may be needed on day 16.

The example liquid chemical use listed in Table 8-3 and charted in Figure 8-1 illustrates how this information can be used. This treatment plant has two 12,000-gal storage tanks. Suppose the operator wishes to keep them full by ordering a truckload (5,000 gal) whenever there is room in the tanks. So, using Figure 8-1, knowing how often the load will be ordered needs to be determined. Using the trendline, the time to use 5,000 gal (between 15,000 and 10,000) is about 8 days (day 5 to day 13). Even when a new load of chemical is ordered, there is about 28 days of storage. This may be more than is desirable.

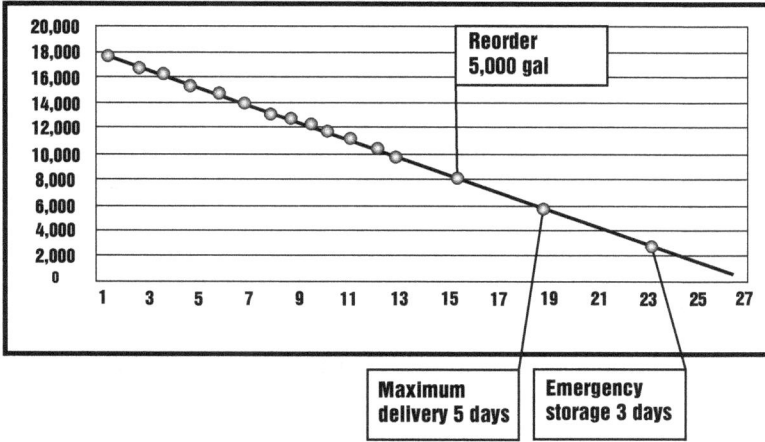

Figure 8-1   Chemical usage deducted from storage (example)

This information can also be used to estimate the time needed to empty a tank (perhaps for maintenance). As an example, a tank has 6,000 gal in storage. Figure 8-1 shows that 6,000 gal is used (say, 16,000 gal to 10,000 gal) in about 10 days (day 3 to day 13). This information can be very helpful when scheduling maintenance.

Chemical Feed in Dilution Tank Systems (Two-Tank Systems)

These types of liquid chemical feed systems use a second tank (dilution tank). Bulk storage tanks may be distant from the chemical feed pump location, there may be concerns about safety of the bulk chemical, or it may be necessary to dilute the chemical for accurate feed or to aid in mixing with process water. Whatever the reason, feed calculations are the same as those given for chemical feed of solutions calculators (c8-10 through c8-13). The only difference is that the percent solution strength is less than for bulk chemicals.

The difference is that the bulk chemical is first diluted prior to feeding. The percent strength of the liquid bulk chemical must be known to accurately determine the percent strength of the diluted solution. Dilution is accomplished by mixing a known volume of the liquid bulk chemical with a known volume of water. The resultant diluted liquid chemical is then fed to the process water using a chemical metering pump.

The easiest way of making diluted solutions is to use a fixed ratio of bulk chemical to dilution water (c8-18).

| c8-18 | % strength of diluted solution | = | % strength of bulk chemical solution | × | Dilution ratio (volume bulk chemical/total volume of diluted solution) |

The conservation of mass relationship may be used to determine the amount of bulk chemical used to make a specified amount of a certain strength diluted solution. This relationship can also be used to determine the amount of the bulk solution used to make an amount of diluted solution. This is based on the fact that the amount of pure chemical is the same for both strength solutions.

| c8-19 | % strength of diluted solution | × | Volume or feed rate of diluted solution | × | % strength of bulk chemical solution | ÷ | Volume or feed rate of chemical solution |

Example 8-8   Five gal of bulk zinc orthophosphate (ZOP) at 14 percent strength is diluted with 50 gal of water. What is the percent strength of the diluted solution?
Use c8-18 percent strength of diluted solution = 14 × 5/55 = 1.3%
Dilution ratio is

volume of bulk chemical/(volume of bulk chemical + volume of water)= 5/55

The percent strength of the diluted solution is used in the calculator for chemical feed of solutions. Chemical usage of the bulk chemical is then calculated back depending on the usage of the diluted chemical, and the frequency that diluted batches must be prepared. Example 8-9 illustrates this.

Example 8-9   Calculate the chemical feed rate in mL/min for ZOP prepared as in Example 8-8. Determine the daily bulk chemical usage (gpd) of this chemical.

Plant flow =	0.14 mgd
ZOP dosage =	3.0 mg/L
ZOP diluted solution strength =	1.3%
ZOP density =	8.34 lb/gal*
ZOP bulk chemical solution strength =	14%

*Density of dilute solutions less than 2 percent strength is assumed to be the same as water. If the specific gravity can be measured, this value can be used, and then calculator c8-9 or c8-10 would be selected.

Units are US, so use calculator c8-10.

$$\text{Chemical feed rate (mL/min)} = 2{,}192 \times \overset{3.0}{\text{Chemical dosage (mg/L)}} \times \overset{0.14}{\text{Water flow (mgd)}} \div \overset{1.3}{\text{Solution strength (\%)}} \div \overset{8.34}{\text{Density of chemical (lb/gal)}}$$

Feed rate of diluted solution (mL/min)
    = 2,192 × 3.0 × 0.14 ÷ 1.3 ÷ 8.34

= 84.9 mL/min

At this feed rate, what is the chemical use of the bulk chemical (gpd)? Use the mass balance ratio c8-19.

$$\overset{1.3}{\underset{\text{solution}}{\substack{\% \\ \text{strength} \\ \text{of diluted}}}} \times \overset{84.9}{\underset{\text{solution}}{\substack{\text{Volume or} \\ \text{feed rate} \\ \text{of diluted}}}} = \overset{14}{\underset{\text{solution}}{\substack{\% \text{ strength} \\ \text{of bulk} \\ \text{chemical}}}} \times \overset{?}{\underset{\text{solution}}{\substack{\text{Volume or} \\ \text{feed rate of} \\ \text{bulk chemical}}}}$$

1.3% × 84.9 mL/min = 14% strength × ?

1.3 × 84.9/14 = 7.9 mL/min bulk solution

Use conversion from mL/min to gpd from Table 8-2.

mL/min ÷ 2.62 = gpd

7.9 mL/min ÷ 2.62 = **3 gpd** of bulk chemical

If the chemical is delivered in 55-gal drums, the drum will last

55 gal/3 gpd = 18 d

Appendix A

Chemical Specification Example Language

The following are examples of specification language for consideration in contracts between "Contractor" and the "Water Department."*

- *All deliveries shall be made by truck between the hours of 8:00 am and 4:30 pm on regular working days of the Water Department, F.O.B. to the following address:*
- *Product shall conform with ANSI/AWWA _____ . Affidavit of compliance must be furnished by supplier. Product must also be certified as meeting ANSI/NSF Standard 60. Documentation of this certification must be provided by supplier. The certifying agency (NSF, UL, CSA, or others) will be notified by the Water Department in the event any certified product arrives at the treatment facility and appears to be contaminated.*
- *Products packaged in bags must be palletized and shrink wrapped. Products in drums must be on pallets.*
- *Product must be delivered in bin/tote containers only—no bulk deliveries will be accepted at this location. Bins/totes must be capable of being lifted from either side with a 4,000-pound capacity forklift.*
- *Pack in multi-wall paper bags of 50 pounds each (net weight) with bags shrink wrapped and palletized.*
- *All products supplied in bag, drum, or bin/tote must have container clearly marked with product designation, name of chemical manufacturer, net weight, and manufacturing lot number. Any product without this labeling may be rejected with vendor to pay freight for return.*
- *The supplier must either use a delivery vehicle dedicated to transporting the product or must supply a washout ticket*

---

* From MacPhee et al. 2002.

111

with information indicating prior material shipped in this vehicle.

- *Delivery vehicles must be properly labeled and must comply with US Department of Transportation (DOT) specifications.*
- *The Water Department has determined that in order to satisfy its Hazard Communication Plan, waste disposal efforts, and regulatory requirements, the Contractor must furnish all available MSDS documentation. This documentation must be provided with the initial delivery, plus after a change in composition of the product, a change in the manufacturing process, a change in supplier, or a change in labeling/ packaging (e.g., change in brand name) of the product.*
- *Delivery personnel shall wear appropriate personnel protective equipment when off-loading product.*
- *Driver will be required to remain at the water treatment plant during unloading. Material may not be off-loaded until authorized by the Water Department.*
- *Product must be delivered to the Water Department in a self-unloading pneumatic tank truck capable of pumping against 25 feet of headloss.*
- *Contractor shall make deliveries in single-unit cargo trailers of 4,500-gallon capacity. The trailer shall be equipped with an appropriate air compressor and pumping system to effectively transfer chemical into the Water Department's storage tanks. The cargo trailer transfer system shall be considered the primary transfer system for off-loading this chemical, with any Water Department–owned transfer system considered as a secondary or backup system.*
- *Contractor shall provide a 2-inch female cam-lock coupling that will be compatible with the Water Department's 2-inch male cam-lock coupling.*
- *Each shipment must contain bill of lading identifying source of supply and a certificate of analysis or "C of A" of the material being shipped by the manufacturer. Information in the C of A must include ....*
- *Every delivery of _____ will be inspected and evaluated by Water Department personnel using tests and test methods chosen at the sole discretion of the Water Department. Each delivery must be approved by Water Department personnel prior to unloading. Material not meeting specifications will be rejected and must be removed at the Contractor's expense.*

- *The bidder shall provide, by attachment to the proposal, a typical analysis of the product, which shall include analysis of the following constituents and parameters: .... The Contractor shall submit such a typical analysis throughout the term of the contract on an annual basis or at the request of the Water Department.*
- *Upon request, the bidder may be required to provide a typical sample of not less than one liter of product.*
- *Liquid alum sulfate (alum) shall be manufactured from aluminum hydrate $(Al(OH)_3)$ and virgin sulfuric acid in a standard digestion process. Alum manufactured from waste products of other processes shall not be accepted. The successful bidder shall submit affidavits indicating compliance with NSF Standard 60 and AWWA Standard B403 (current edition) and with the requirement regarding raw materials specified in this paragraph. The affidavit concerning raw material processes must be submitted annually.*
- *Ton containers or 150-pound cylinders of liquid chlorine, which cannot be opened with reasonable force by plant personnel, will be returned to the supplier for exchange at no additional charge.*
- *The Water Department reserves the right to reject any shipment that does not include the proper documentation or that does not match the Water Department's screening tests. In the event a delivery is rejected, the Contractor shall provide a replacement shipment meeting the requirements of this specification within 24 hours.*

Appendix B

Conversion Factors[*]

Conversion of US Customary Units

Linear Measurement
fathoms × 6 = feet (ft)
feet (ft) × 12 = inches (in.)
inches (in.) × 0.0833 = feet (ft)
miles (mi) × 5,280 = feet (ft)
yards (yd) × 3 = feet (ft)
yards (yd) × 36 = inches (in.)

Circular Measurement
degrees (angle) × 60 = minutes (angle)
degrees (angle) × 0.01745 = radians

Area Measurement
acres × 43,560 = square feet (ft^2)
square feet (ft^2) × 144 = square inches $(in.^2)$
square inches $(in.^2)$ × 0.00695 = square feet (ft^2)
square miles (mi^2) × 640 = acres
square miles (mi^2) × 27,878,400 = square feet (ft^2)
square miles (mi^2) × 3,098,000 = square yards (yd^2)
square yards (yd^2) × 9 = square feet (ft^2)

Volume Measurement
acre-feet (acre-ft) × 43,560 = cubic feet (ft^3)
acre-feet (acre-ft) × 325,851 = gallons (gal)
barrels petroleum (bo) × 42 = gallons (gal)
board foot (fbm)　 = 144 square inches × 1 inch
cubic feet (ft^3) × 1,728 = cubic inches $(in.^3)$
cubic feet (ft^3) × 7.48052 = gallons (gal)
cubic feet (ft^3) × 29.92 = quarts (qt)

---

* From Lauer et al. 2004, 31–41.

cubic feet (ft^3) × 59.84 = pints (pt)
cubic feet (ft^3) × 0.000023 = acre feet (acre-ft)
cubic inches (in.^{3}) × 0.00433 = gallons (gal)
cubic inches (in.^{3}) × 0.00058 = cubic feet (ft^3)
drops × 60 = teaspoons (tsp)
gallons (gal) × 0.1337 = cubic feet (ft^3)
gallons (gal) × 231 = cubic inches (in.^{3})
gallons (gal) × 4 = quarts (qt)
gallons (gal) × 8 = pints (pt)
gallons, US × 0.83267 = gallons, Imperial
gallons (gal) × 0.00000308 = acre-feet (acre-ft)
gallons (gal) × 0.0238 = barrels (petroleum) (bo)
gallons, Imperial × 1.20095 = gallons, US
pints (pt) × 2 = quarts (qt)
quarts (qt) × 4 = gallons (gal)
quarts (qt) × 57.75 = cubic inches (in.^{3})

Pressure Measurement

atmospheres × 29.92 = inches of mercury
atmospheres × 33.90 = feet of water
atmospheres × 14.70 = pounds per square inch (lb/in.^{2})
feet of water × 0.8826 = inches of mercury
feet of water × 0.02950 = atmospheres
feet of water × 0.4335 = pounds per square inch (lb/in.^{2})
feet of water × 62.43 = pounds per square foot (lb/ft^2)
feet of water × 0.8876 = inches of mercury
inches of mercury × 1.133 = feet of water
inches of mercury × 0.03342 = atmospheres
inches of mercury × 0.4912 = pounds per square inch (lb/in.^{2})
inches of water × 0.002458 = atmospheres
inches of water × 0.07355 = inches of mercury
inches of water × 0.03613 = pounds per square inch (lb/in.^{2})
pounds/square in. (lb/in.^{2}) × 144 = pounds per square foot (lb/ft^2)
pounds/square foot (lb/ft^2) × .00694 = pounds per square inch (lb/in.^{2})
pounds/square in. (lb/in.^{2}) × 2.307 = feet of water
pounds/square inch (lb/in.^{2}) × 2.036 = inches of mercury
pounds/square inch (lb/in.^{2}) × 27.70 = inches of water

Weight Measurement

cubic feet of ice × 57.2 = pounds (lb)
cubic feet of water (50°F) × 62.4 = pounds of water
cubic inches of water × 0.036 = pounds of water
gallons of water (50°F) × 8.3453 = pounds of water
milligrams/liter (mg/L) × 0.0584 = grains per gallon (US) (gpg)

milligrams/liter (mg/L) × 0.07016 = grains per gallon (UK) (Imp)
milligrams/liter (mg/L) × 8.345 = pounds per million gallons
    (lb/mil gal)
ounces (oz) × 437.5 = grains (gr)
parts per million (ppm) × 1 = milligrams per liter (mg/L) (for normal
    water applications)
grains per gallon (gpg) × 17.118 = parts per million (ppm)
grains per gallon (gpg) × 142.86 = pounds per million gallons
    (lb/mil gal)
percent solution × 10,000 = milligrams per liter (mg/L)
pounds (lb) × 16 = ounces (oz)
pounds (lb) × 7,000 = grains (gr)
pounds (lb) × 0.0005 = tons (short)
pounds/cubic inch (lb/in.^{3}) × 1,728 = pounds per cubic foot (lb/ft^3)
pounds of water × 0.0160 = cubic feet (ft^3)
pounds of water × 27.68 = cubic inches (in.^{3})
pounds of water × 0.1198 = gallons (gal)
tons (short) × 2,000 = pounds (lb)
tons (short) × 0.89287 = tons (long)
tons (long) × 2,240 = pounds (lb)
cubic feet air (at 60°F and 29.92 in. mercury) × 0.0763 = pounds (lb)

Flow Measurement

barrels per hour petroleum (bo/hr) × 0.70 = gallons per minute (gpm)
acre-feet/minute (acre-ft/min) × 325,853 = gallons per minute (gpm)
acre-feet/minute (acre-ft/min) × 726 = cubic feet per second (ft^3/sec)
cubic feet/minute (ft^3/min) × 0.1247 = gallons per second (gps)
cubic feet/minute (ft^3/min) × 62.43 = pounds of water per minute
cubic feet/second (ft^3/sec) × 448.831 = gallons per minute (gpm)
cubic feet/second (ft^3/sec) × 0.646317 = million gallons per day (mgd)
cubic feet/second (ft^3/sec) × 1.984 = acre-feet per day (acre-ft/day)
gallons/minute (gpm) × 1,440 = gallons per day (gpd)
gallons/minute (gpm) × 0.00144 = million gallons per day (mgd)
gallons/minute (gpm) × 0.00223 = cubic feet per second (ft^3/sec)
gallons/minute (gpm) × 0.1337 = cubic feet per minute (ft^3/min)
gallons/minute (gpm) × 8.0208 = cubic feet per hour (ft^3/hr)
gallons/minute (gpm × 0.00442 = acre-feet per day (acre-ft/day)
gallons/minute (gpm) × 1.43 = barrels (42 petroleum gal) per hour
    (bo/day)
gallons water/minute × 6.0086 = tons of water per 24 hours
million gallons/day (mgd) × 1.54723 = cubic feet per second (ft^3/sec)
million gallons/day (mgd) × 92.82 = cubic feet per minute (ft^3/min)
million gallons/day (mgd) × 694.4 = gallons per minute (gpm)
million gallons/day (mgd) × 3.07 = acre-feet per day (acre-ft/day)
pounds of water/minute × 0.000267 = cubic feet per second (ft^3/sec)

Work Measurement

British thermal units (Btu) × 778.2 = foot-pounds (ft-lb)
British thermal units (Btu) × 0.000393 = horsepower-hours (hp·hr)
British thermal units (Btu) × 0.000293 = kilowatt-hours (kW·hr)
foot-pounds (ft-lb) × 0.001286 = British thermal units (Btu)
foot-pounds (ft-lb) × 0.000000505 = horsepower-hours (hp·hr)
foot-pounds (ft-lb) × 0.000000377 = kilowatt-hours (kW·hr)
horsepower-hours (hp·hr) × 2,547 = British thermal units (Btu)
horsepower-hours (hp·hr) × 0.7457 = kilowatt-hours (kW·hr)
kilowatt-hours (kW·hr) × 3,412 = British thermal units (Btu)
kilowatt-hours (kW·hr) × 1.341 = horsepower-hours (hp·hr)

Power Measurement

boiler horsepower × 33,480 = British thermal units per hour (Btu/hr)
boiler horsepower × 9.8 = kilowatts (kW)
British thermal units/second (Btu/sec) × 1.0551 = kilowatts (kW)
British thermal units/minute (Btu/min) × 12.96 = foot-pounds per second (ft-lb/sec)
British thermal units/minute (Btu/min) × 0.02356 = horsepower (hp)
British thermal units/minute (Btu/min) × 0.01757 = kilowatts (kW)
British thermal units/hour (Btu/hr) × 0.293 = watts (W)
British thermal units/hour (Btu/hr) × 12.96 = foot-pounds per minute (ft-lb/min)
British thermal units/hour (Btu/hr) × 0.00039 = horsepower (hp)
foot-pounds per second (ft-lb/sec) × 0.0771 = British thermal units per minute (Btu/min)
foot-pounds per second (ft-lb/sec) × 0.001818 = horsepower (hp)
foot-pounds per second (ft-lb/sec) × 0.001356 = kilowatts (kW)
foot-pounds per minute (ft-lb/min) × 0.0000303 = horsepower (hp)
foot-pounds per minute (ft-lb/min) × 0.0000226 = kilowatts (kW)
horsepower (hp) × 42.44 = British thermal units per minute (Btu/min)
horsepower (hp) × 33,000 = foot-pounds per minute (ft-lb/min)
horsepower (hp) × 550 = foot-pounds per second (ft-lb/sec)
horsepower (hp) × 1,980,000 = foot-pounds per hour (ft-lb/hr)
horsepower (hp) × 0.7457 = kilowatts (kW)
horsepower (hp) × 745.7 = watts (W)
kilowatts (kW) × 0.9478 = British thermal units per second (Btu/sec)
kilowatts (kW) × 56.87 = British thermal units per minute (Btu/min)
kilowatts (kW) × 3,413 = British thermal units per hour (Btu/hr)
kilowatts (kW) × 44,250 = foot-pounds per minute (ft-lb/min)
kilowatts (kW) × 737.6 = foot-pounds per second (ft-lb/sec)
kilowatts (kW) × 1.341 = horsepower (hp)
tons of refrigeration (US) × 288,000 = British thermal units per 24 hours
watts (W) × 0.05692 = British thermal units per minute (Btu/min)

watts (W) × 0.7376 = foot-pounds (force) per second (ft-lb/sec)
watts (W) × 44.26 = foot-pounds per minute (ft-lb/min)
watts (W) × 0.001341 = horsepower (hp)

Velocity Measurement

feet/minute (ft/min) × 0.01667 = feet per second (ft/sec)
feet/minute (ft/min) × 0.01136 = miles per hour (mph)
feet/second (ft/sec) × 0.6818 = miles per hour (mph)
miles/hour (mph) × 88 = feet per minute (ft/min)
miles/hour (mph) × 1.467 = feet per second (ft/sec)

Miscellaneous

grade: 1 percent (or 0.01) = 1 foot per 100 feet

Metric Conversions

Linear Measurement

inch (in.) × 25.4 = millimeters (mm)
inch (in.) × 2.54 = centimeters (cm)
foot (ft) × 304.8 = millimeters (mm)
foot (ft) × 30.48 = centimeters (cm)
foot (ft) × 0.3048 = meters (m)
yard (yd) × 0.9144 = meters (m)
mile (mi) × 1,609.3 = meters (m)
mile (mi) × 1.6093 = kilometers (km)
millimeter (mm) × 0.03937 = inches (in.)
centimeter (cm) × 0.3937 = inches (in.)
meter (m) × 39.3701 = inches (in.)
meter (m) × 3.2808 = feet (ft)
meter (m) × 1.0936 = yards (yd)
kilometer (km) × 0.6214 = miles (mi)

Area Measurement

square meter (m^2) × 10,000 = square centimeters (cm^2)
hectare (ha) × 10,000 = square meters (m^2)
square inch (in.^{2}) × 6.4516 = square centimeters (cm^2)
square foot (ft^2) × 0.092903 = square meters (m^2)
square yard (yd^2) × 0.8361 = square meters (m^2)
acre × 0.004047 = square kilometers (km^2)
acre × 0.4047 = hectares (ha)
square mile (mi^2) × 2.59 = square kilometers (km^2)
square centimeter (cm^2) × 0.16 = square inches (in.^{2})
square meters (m^2) × 10.7639 = square feet (ft^2)

square meters (m^2) × 1.1960 = square yards (yd^2)
hectare (ha) × 2.471 = acres
square kilometer (km^2) × 247.1054 = acres
square kilometer (km^2) × 0.3861 = square miles (mi^2)

Volume Measurement

cubic inch (in.^{3}) × 16.3871 = cubic centimeters (cm^3)
cubic foot (ft^3) × 28,317 = cubic centimeters (cm^3)
cubic foot (ft^3) × 0.028317 = cubic meters (m^3)
cubic foot (ft^3) × 28.317 = liters (L)
cubic yard (yd^3) × 0.7646 = cubic meters (m^3)
acre foot (acre-ft) × 1,233.4 = cubic meters (m^3)
ounce (US fluid) (oz) × 0.029573 = liters (L)
quart (liquid) (qt) × 946.9 = milliliters (mL)
quart (liquid) (qt) × 0.9463 = liters (L)
gallon (gal) × 3.7854 = liters (L)
gallon (gal) × 0.0037854 = cubic meters (m^3)
peck (pk) × 0.881 = decaliters (dL)
bushel (bu) × 0.3524 = hectoliters (hL)
cubic centimeters (cm^3) × 0.061 = cubic inches (in.^{3})
cubic meter (m^3) × 35.3183 = cubic feet (ft^3)
cubic meter (m^3) × 1.3079 = cubic yards (yd^3)
cubic meter (m^3) × 264.2 = gallons (gal)
cubic meter (m^3) × 0.000811 = acre-feet (acre-ft)
liter (L) × 1.0567 = quart (liquid) (qt)
liter (L) × 0.264 = gallons (gal)
liter (L) × 0.0353 = cubic feet (ft^3)
decaliter (dL) × 2.6417 = gallons (gal)
decaliter (dL) × 1.135 = pecks (pk)
hectoliter (hL) × 3.531 = cubic feet (ft^3)
hectoliter (hL) × 2.84 = bushels (bu)
hectoliter (hL) × 0.131 = cubic yards (yd^3)
hectoliter (hL) × 26.42 = gallons (gal)

Pressure Measurement

pound/square inch (psi) × 6.8948 = kilopascals (kPa)
pound/square inch (psi) × 6,894 = pascals (Pa)
pound/square inch (psi) × 0.070307 = kilograms/square centimeter (kg/cm^2)
pound/square foot (lb/ft^2) × 47.8803 = pascals (Pa)
pound/square foot (lb/ft^2) × 0.000488 = kilograms/square centimeter (kg/cm^2)
pound/square foot (lb/ft^2) × 4.8824 = kilograms/square meter (kg/m^2)
inches of mercury × 3,386.4 = pascals (Pa)

inches of water × 248.84 = pascals (Pa)
bar × 100,000 = newtons per square meter (N/m^2)
pascals (Pa) × 1 = newtons per square meter (N/m^2)
pascals (Pa) × 0.000145 = pounds/square inch (psi)
kilopascals (kPa) × 0.145 = pounds/square inch (psi)
pascals (Pa) × 0.000296 = inches of mercury (at 60°F)
kilogram/square centimeter (kg/cm^2) × 14.22 = pounds/square inch (psi)
kilogram/square centimeter (kg/cm^2) × 28.959 = inches of mercury
    (at 60°F)
kilogram/square meter
    (kg/m^2) × 0.2048 = pounds per square foot (lb/ft^2)
centimeters of mercury × 0.4461 = feet of water

Weight Measurement

ounce (oz) × 28.3495 = grams (g)
pound (lb) × 453.59 = grams (g)
pound (lb) × 0.4536 = kilograms (kg)
ton (short) × 0.9072 = megagrams (metric ton)
pounds/cubic foot (lb/ft^3) × 16.02 = grams per liter (g/L)
pounds/million gallons (lb/mil gal) × 0.1198 = grams per cubic
    meter (g/m^3)
gram (g) × 15.4324 = grains (gr)
gram (g) × 0.0353 = ounces (oz)
gram (g) × 0.0022 = pounds (lb)
kilograms (kg) × 2.2046 = pounds (lb)
kilograms (kg) × 0.0011 = tons (short)
megagram (metric ton) × 1.1023 = tons (short)
grams/liter (g/L) × 0.0624 = pounds per cubic foot (lb/ft^3)
grams/cubic meter (g/m^3) × 8.3454 = pounds/million gallons
    (lb/mil gal)

Flow Measurement

gallons/second (gps) × 3.785 = liters per second (L/sec)
gallons/minute (gpm) × 0.00006308 = cubic meters per second (m^3/sec)
gallons/minute (gpm) × 0.06308 = liters per second (L/sec)
gallons/hour (gph) × 0.003785 = cubic meters per hour (m^3/hr)
gallons/day (gpd) × 0.000003785 = million liters per day (ML/day)
gallons/day (gpd) × 0.003785 = cubic meters per day (m^3/day)
cubic feet/second (ft^3/sec) × 0.028317 = cubic meters per second
    (m^3/sec)
cubic feet/second (ft^3/sec) × 1,699 = liters per minute (L/min)
cubic feet/minute (ft^3/min) × 472 = cubic centimeters/second (cm^3/sec)
cubic feet/minute (ft^3/min) × 0.472 = liters per second (L/sec)
cubic feet/minute (ft^3/min) × 1.6990 = cubic meters per hour (m^3/hr)
million gallons/day (mgd) × 43.8126 = liters per second (L/sec)

million gallons/day (mgd) × 3,785 = cubic meters per day (m³/day)
million gallons/day (mgd) × 0.043813 = cubic meters per second
   (m³/sec)
gallons/square foot (gal/ft²) × 40.74 = liters per square meter (L/m²)
gallons/acre/day (gal/acre/day) × 0.0094 = cubic meters/hectare/day
   (m³/ha/day)
gallons/square foot/day (gal/ft²/day) × 0.0407 = cubic meters/square
   meter/day (m³/m²/day)
gallons/square foot/day (gal/ft²/day) × 0.0283 = liters/square meter/
   min (L/m²/m)
gallons/square foot/minute (gal/ft²/min) × 2.444 = cubic meters/square
   meter/hour (m³/m²/hr) = m/hr
gallons/square foot/minute (gal/ft²/min) × 0.679 = liters/square meter/
   second (L/m²/sec)
gallons/square foot/minute (gal/ft²/min) × 40.7458 = liters/square
   meter/minute (L/m²/min)
gallons/capita/day (gpcd) × 3.785 = liters/day/capita (L/d/capita)
liters/second (L/sec) × 22,824.5 = gallons per day (gpd)
liters/second (L/sec) × 0.0228 = million gallons per day (mgd)
liters/second (L/sec) × 15.8508 = gallons per minute (gpm)
liters/second (L/sec) × 2.119 = cubic feet per minute (ft³/min)
liters/minute (L/min) × 0.0005886 = cubic feet per second (ft³/sec)
cubic centimeters/second (cm³/sec) × 0.0021 = cubic feet per minute
   (ft³/min)
cubic meters/second (m³/sec) × 35.3147 = cubic feet per second (ft³/sec)
cubic meters/second (m³/sec) × 22.8245 = million gallons per day (mgd)
cubic meters/second (m³/sec) × 15,850.3 = gallons per minute (gpm)
cubic meters/hour (m³/hr) × 0.5886 = cubic feet per minute (ft³/min)
cubic meters/hour (m³/hr) × 4.403 = gallons per minute (gpm)
cubic meters/day (m³/day) × 264.1720 = gallons per day (gpd)
cubic meters/day (m³/day) × 0.00026417 = million gallons per day (mgd)
cubic meters/hectare/day (m³/ha/day) × 106.9064 = gallons per acre
   per day (gal/acre/day)
cubic meters/square meter/day (m³/m²/day) × 24.5424 = gallons/square
   foot/day (gal/ft²/day)
liters/square meter/minute (L/m²/min) × 0.0245 = gallons/square foot/
   minute (gal/ft²/min)
liters/square meter/minute (L/m²/min) × 35.3420 = gallons/square
   foot/day (gal/ft²/day)

Work, Heat, and Energy Measurements
British thermal units (Btu) × 1.0551 = kilojoules (kJ)
British thermal units (Btu) × 0.2520 = kilogram-calories (kg-cal)
foot-pound (force) (ft-lb) × 1.3558 = joules (J)
horsepower-hour (hp·hr) × 2.6845 = megajoules (MJ)
watt-second (W-sec) × 1.000 = joules (J)

watt-hour (W·hr) × 3.600 = kilojoules (kJ)

kilowatt-hour (kW·hr) × 3,600 = kilojoules (kJ)

kilowatt-hour (kW·hr) × 3,600,000 = joules (J)

British thermal units per pound (Btu/lb) × 0.5555 = kilogram-calories per kilogram (kg-cal/kg)

British thermal units per cubic foot (Btu/ft^3) × 8.8987 = kilogram-calories/cubic meter (kg-cal/m^3)

kilojoule (kJ) × 0.9478 = British thermal units (Btu)

kilojoule (kJ) × 0.00027778 = kilowatt-hours (kW·hr)

kilojoule (kJ) × 0.2778 = watt-hours (W·hr)

joule (J) × 0.7376 = foot-pounds (ft-lb)

joule (J) × 1.0000 = watt-seconds (W-sec)

joule (J) × 0.2399 = calories (cal)

megajoule (MJ) × 0.3725 = horsepower-hour (hp·hr)

kilogram-calories (kg-cal) × 3.9685 = British thermal units (Btu)

kilogram-calories per kilogram (kg-cal/kg) × 1.8000 = British thermal units per pound (Btu/lb)

kilogram-calories per liter (kg-cal/L) × 112.37 = British thermal units per cubic foot (Btu/ft^3)

kilogram-calories/cubic meter (kg-cal/m^3) × 0.1124 = British thermal units per cubic foot (Btu/ft^3)

Velocity, Acceleration, and Force Measurements

feet per minute (ft/min) × 18.2880 = meters per hour (m/hr)

feet per hour (ft/hr) × 0.3048 = meters per hour (m/hr)

miles per hour (mph) × 44.7 = centimeters per second (cm/sec)

miles per hour (mph) × 26.82 = meters per minute (m/min)

miles per hour (mph) × 1.609 = kilometers per hour (km/hr)

feet/second/second (ft/sec^2) × 0.3048 = meters/second/second (m/sec^2)

inches/second/second (in./sec^2) × 0.0254 = meters/second/second (m/sec^2)

pound-force (lbf) × 4.44482 = newtons (N)

centimeters/second (cm/sec) × 0.0224 = miles per hour (mph)

meters/second (m/sec) × 3.2808 = feet per second (ft/sec)

meters/minute (m/min) × 0.0373 = miles per hour (mph)

meters per hour (m/hr) × 0.0547 = feet per minute (ft/min)

meters per hour (m/hr) × 3.2808 = feet per hour (ft/hr)

kilometers/second (km/sec) × 2,236.9 = miles per hour (mph)

kilometers/hour (km/hr) × 0.0103 = miles per min (mpm)

meters/second/second (m/sec^2) × 3.2808 = feet/second/second (ft/sec^2)

meters/second/second (m/sec^2) × 39.3701 = inches/second/second (in./sec^2)

newtons (N) × 0.2248 = pounds force (lbf)

Appendix C
Density Tables for Common Liquid Chemicals

Table C-1 Properties of liquid alum

Specific Gravity, g/mL	lb/gal	% Al$_2$O$_3$	Equivalent % Dry Alum*	Dry Alum per Gallon Solution, lb	Dry Alum per Liter Solution, g
1.0069	8.40	0.19	1.12	0.09	11.277
1.0140	8.46	0.39	2.29	0.19	23.221
1.0211	8.52	0.59	3.47	0.30	35.432
1.0284	8.58	0.80	4.71	0.40	48.438
1.0357	8.64	1.01	5.94	0.51	61.521
1.0432	8.70	1.22	7.18	0.62	74.902
1.0507	8.76	1.43	8.41	0.74	88.364
1.0584	8.83	1.64	9.65	0.85	102.136
1.0662	8.89	1.85	10.88	0.97	116.003
1.0741	8.96	2.07	12.18	1.09	130.825
1.0821	9.02	2.28	13.41	1.21	145.110
1.0902	9.09	2.50	14.71	1.34	160.368
1.0985	9.16	2.72	16.00	1.47	175.760
1.1069	9.23	2.93	17.24	1.59	190.830
1.1154	9.30	3.15	18.53	1.72	206.684
1.1240	9.37	3.38	19.88	1.86	223.451
1.1328	9.45	3.60	21.18	2.00	239.927
1.1417	9.52	3.82	22.47	2.14	256.540
1.1508	9.60	4.04	23.76	2.28	273.430
1.1600	9.67	4.27	25.12	2.43	291.392
1.1694	9.75	4.50	26.47	2.58	309.540
1.1789	9.83	4.73	27.82	2.74	327.970
1.1885	9.91	4.96	29.18	2.89	346.804
1.1983	9.99	5.19	30.53	3.05	365.841
1.2083	10.08	5.43	31.94	3.22	385.931
1.2185	10.16	5.67	33.35	3.39	406.370
1.2288	10.25	5.91	34.76	3.56	427.131
1.2393	10.34	6.16	36.24	3.74	449.122
1.2500	10.43	6.42	37.76	3.93	472.000
1.2609	10.52	6.67	39.24	4.12	494.777
1.2719	10.61	6.91	40.65	4.31	517.027
1.2832	10.70	7.16	42.12	4.51	540.484
1.2946	10.80	7.40	43.53	4.71	563.539
1.3063	10.89	7.66	45.06	4.91	588.619
1.3182	10.99	7.92	46.59	5.12	614.149
1.3303	11.09	8.19	48.18	5.34	640.938
1.3426	11.20	8.46	49.76	5.57	668.078
1.3551	11.30	8.74	51.41	5.81	696.657
1.3679	11.41	9.01	53.00	6.05	724.987

* 17% Al$_2$O$_3$ in dry alum + 0.03% free Al$_2$O$_3$.

Source: Lauer et al. 2004, 189–190.

Table C-2   Densities and weight equivalents of standard alum solutions (with temperature correction)

Specific Gravity (60°F)	Lb/US Gallon	% Free Al$_2$O$_3$ = 0.05		% Free Al$_2$O$_3$ = 0.10		% Free Al$_2$O$_3$ = 0.15		% Free Al$_2$O$_3$ = 0.20	
		% Total Al$_2$O$_3$	% Dry Alum	% Total Al$_2$O$_3$	% Dry Alum	% Total Al$_2$O$_3$	% Dry Alum	% Total Al$_2$O$_3$	% Dry Alum
1.3110	10.940							8.00	46.6
1.3120	10.948							8.02	46.7
1.3130	10.956					8.00	46.6	8.03	46.8
1.3140	10.965					8.01	46.7	8.05	46.9
1.3150	10.973					8.03	46.7	8.06	46.9
1.3160	10.981			8.01	46.6	8.04	46.8	8.07	47.0
1.3170	10.990			8.02	46.7	8.05	46.9	8.09	47.1
1.3180	10.998	8.00	46.6	8.03	46.8	8.07	47.0	8.10	47.2
1.3190	11.006	8.01	46.7	8.05	46.9	8.08	47.1	8.11	47.3
1.3200	11.015	8.03	46.7	8.06	46.9	8.09	47.1	8.13	47.3
1.3210	11.023	8.04	46.8	8.07	47.0	8.11	47.2	8.14	47.4
1.3220	11.031	8.05	46.9	8.09	47.1	8.12	47.3	8.16	47.5
1.3230	11.040	8.07	47.0	8.10	47.2	8.13	47.4	8.17	47.6
1.3240	11.048	8.08	47.1	8.12	47.3	8.15	47.5	8.18	47.7
1.3250	11.056	8.09	47.1	8.13	47.3	8.16	47.5	8.20	47.7
1.3260	11.065	8.11	47.2	8.14	47.4	8.18	47.6	8.21	47.8
1.3270	11.073	8.12	47.3	8.16	47.5	8.19	47.7	8.22	47.9
1.3280	11.081	8.14	47.4	8.17	47.6	8.20	47.8	8.24	48.0
1.3290	11.090	8.15	47.5	8.18	47.7	8.22	47.9	8.25	48.1
1.3300	11.098	8.16	47.5	8.20	47.7	8.23	47.9	8.27	48.1
1.3310	11.107	8.18	47.6	8.21	47.8	8.24	48.0	8.28	48.2

Table continues next page.

Table C-2   Densities and weight equivalents of standard alum solutions (with temperature correction) (continued)

Specific Gravity (60°F)	Lb/US Gallon	% Free Al$_2$O$_3$ = 0.05		% Free Al$_2$O$_3$ = 0.10		% Free Al$_2$O$_3$ = 0.15		% Free Al$_2$O$_3$ = 0.20	
		% Total Al$_2$O$_3$	% Dry Alum	% Total Al$_2$O$_3$	% Dry Alum	% Total Al$_2$O$_3$	% Dry Alum	% Total Al$_2$O$_3$	% Dry Alum
1.3320	11.115	8.19	47.7	8.22	47.9	8.26	48.1	8.29	48.3
1.3330	11.123	8.20	47.8	8.24	48.0	8.27	48.2	8.31	48.4
1.3340	11.132	8.22	47.8	8.25	48.1	8.29	48.2	8.32	48.5
1.3350	11.140	8.23	47.9	8.27	48.1	8.30	48.3	8.33	48.5
1.3360	11.148	8.24	48.0	8.28	48.2	8.31	48.4	8.35	48.6
1.3370	11.157	8.26	48.1	8.29	48.3	8.33	48.5	8.36	48.7
1.3380	11.165	8.27	48.2	8.31	48.4	8.34	48.6	8.38	48.8
1.3390	11.173	8.28	48.2	8.32	48.5	8.35	48.6	8.39	48.9
1.3400	11.182	8.30	48.3	8.33	48.5	8.37	48.7	8.40	48.9
1.3410	11.190	8.31	48.4	8.35	48.6	8.38	48.8		
1.3420	11.198	8.33	48.5	8.36	48.7	8.39	48.9		
1.3430	11.207	8.34	48.6	8.37	48.8				
1.3440	11.215	8.35	48.6	8.39	48.8				
1.3450	11.223	8.37	48.7	8.40	48.9				
1.3460	11.232	8.38	48.8						
1.3470	11.240	8.39	48.9						

Adapted from General Chemical Corp., Brighton, Mich.

Table C-3 Temperature corrections for standard alum solutions (specific gravity over 1.28)

Add the correction to the specific gravity reading taken at the corresponding temperature to obtain the specific gravity at 60°F

Temperature, °F	Correction	Temperature, °F	Correction	Temperature, °F	Correction	Temperature, °F	Correction
40	−0.0049	80	0.0049	104	0.0125	128	0.0220
45	−0.0036	81	0.0051	105	0.0129	129	0.0225
50	−0.0022	82	0.0054	106	0.0133	130	0.0230
55	−0.0011	83	0.0058	107	0.0136	131	0.0233
60	0	84	0.0060	108	0.0140	132	0.0238
61	0.0002	85	0.0064	109	0.0144	133	0.0243
62	0.0004	86	0.0066	110	0.0148	134	0.0247
63	0.0006	87	0.0068	111	0.0152	135	0.0252
64	0.0009	88	0.0072	112	0.0156	136	0.0256
65	0.0011	89	0.0075	113	0.0159	137	0.0260
66	0.0014	90	0.0078	114	0.0163	138	0.0265
67	0.0016	91	0.0081	115	0.0166	139	0.0270
68	0.0018	92	0.0084	116	0.0171	140	0.0275
69	0.0020	93	0.0088	117	0.0175	141	0.0279
70	0.0022	94	0.0090	118	0.0178	142	0.0284
71	0.0025	95	0.0094	119	0.0183	143	0.0289

Table continues next page.

Table C-3   Temperature corrections for standard alum solutions (specific gravity over 1.28) (continued)

Add the correction to the specific gravity reading taken at
the corresponding temperature to obtain the specific gravity at 60°F

Temperature, °F	Correction	Temperature, °F	Correction	Temperature, °F	Correction	Temperature, °F	Correction
72	0.0028	96	0.0098	120	0.0187	144	0.0293
73	0.0031	97	0.0101	121	0.0190	145	0.0298
74	0.0033	98	0.0105	122	0.0195	146	0.0303
75	0.0036	99	0.0107	123	0.0199	147	0.0309
76	0.0038	100	0.0111	124	0.0203	148	0.0313
77	0.0040	101	0.0115	125	0.0207	149	0.0318
78	0.0043	102	0.0118	126	0.0212	150	0.0324
79	0.0047	103	0.0122	127	0.0217		

Adapted from General Chemical Corp., Brighton, Mich.

Table C-4   Sodium hypochlorite specific gravity table

Trade % (g available Cl_2/100 mL sol.)	%w/w Cl_2 (g available Cl_2/ 100 g sol.)	%w/vol NaOCl (g NaOCl/ 100 mL sol.)	%w/w NaOCl (g NaOCl/ 100 g sol.)	Specific Gravity
11.00	9.47	11.55	9.95	1.161
12.00	10.22	12.60	10.73	1.174
12.50	10.59	13.12	11.12	1.180
14.00	11.69	14.70	12.27	1.198
15.00	12.40	15.75	13.02	1.210
16.00	13.09	16.80	13.75	1.222
10.00	15.09	19.95	15.84	1.259
19.05	15.12	20.00	15.87	1.260
19.64	15.50	20.62	16.27	1.267
20.00	15.74	21.00	16.52	1.271

Table C-5   Percent hydrofluosilicic acid (H_2SiF_6) vs. density chart 60°C

%H_2SiF_6	Density	%H_2SiF_6	Density
15.0	1.134	23.0	1.206
15.5	1.139	23.5	1.210
16.0	1.143	24.0	1.215
16.5	1.148	24.5	1.219
17.0	1.152	25.0	1.224
17.5	1.156	25.5	1.228
18.0	1.161	26.0	1.232
18.5	1.165	26.5	1.237
19.0	1.170	27.0	1.241
19.5	1.174	27.5	1.246
20.0	1.179	28.0	1.250
20.5	1.183	28.5	1.255
21.0	1.188	29.0	1.259
21.5	1.192	29.5	1.264
22.0	1.197	30.0	1.268
22.5	1.201		
Source: Courtesy of Mosaic Co.			

Table C-6   Densities of pure (salt free) caustic soda solution (volumetric units) at 20°C (68°F)

% NaOH	% Na$_2$O	Specific Gravity 20/4°C	°Baumé Am. Std.	°Twaddell	NaOH (Dry lb/gal)	Total Weight Solution (lb/gal)	NaOH (Dry lb/IG)	Total Weight Solution (lb/IG)	NaOH (Dry g/liter)	Total Weight Solution (g/liter)	NaOH (Dry lb/ft³)	Total Weight Solution (lb/ft³)
1	0.78	1.010	1.4	1.9	0.08	8.41	0.10	10.10	10.1	1,008	0.63	62.90
2	1.55	1.021	2.9	4.1	0.17	8.50	0.20	10.21	20.4	1,019	1.28	63.59
3	2.33	1.032	4.5	6.4	0.26	8.59	0.31	10.32	31.0	1,030	1.93	64.29
4	3.10	1.043	6.0	8.6	0.35	8.68	0.42	10.43	41.8	1,041	2.60	64.99
5	3.88	1.054	7.4	10.8	0.44	8.78	0.53	10.54	52.8	1,052	3.29	65.66
6	4.65	1.065	8.8	13.0	0.53	8.87	0.64	10.65	63.9	1,063	3.99	66.34
7	5.43	1.076	10.2	15.2	0.63	8.96	0.75	10.76	75.3	1,074	4.70	67.03
8	6.20	1.087	11.6	17.4	0.73	9.05	0.87	10.87	86.9	1,085	5.43	67.72
9	6.98	1.098	12.9	19.6	0.82	9.14	0.99	10.98	98.8	1,096	6.17	68.40
10	7.75	1.109	14.2	21.8	0.93	9.24	1.11	11.09	110.9	1,107	6.92	69.09
12	9.30	1.131	16.8	26.2	1.13	9.42	1.36	11.31	135.7	1,129	8.47	70.46
14	10.85	1.153	19.2	30.6	1.35	9.60	1.62	11.53	161.4	1,151	10.08	71.84
16	12.40	1.175	21.6	35.0	1.57	9.79	1.88	11.75	188.0	1,173	11.74	73.21
18	13.95	1.197	23.9	39.4	1.80	9.97	2.16	11.98	215.5	1,195	13.45	74.59
20	15.50	1.219	26.1	43.8	2.04	10.15	2.44	12.19	243.8	1,217	15.22	75.96
22	17.05	1.241	28.2	48.2	2.28	10.34	2.74	12.41	273.0	1,239	17.05	77.33
24	18.60	1.263	30.2	52.6	2.53	10.52	3.04	12.63	303.1	1,261	18.92	78.69
26	20.15	1.285	32.1	57.0	2.79	10.70	3.35	12.85	334.0	1,282	20.85	80.05
28	21.70	1.306	34.0	61.3	3.05	10.88	3.67	13.07	365.8	1,304	22.84	81.40

Table continues next page.

Table C-6   Densities of pure (salt free) caustic soda solution (volumetric units) at 20°C (68°F) (continued)

% NaOH	% Na$_2$O	Specific Gravity 20/4°C	°Baumé Am. Std.	°Twaddell	NaOH (Dry lb/gal)	Total Weight Solution (lb/gal)	NaOH (Dry lb/IG)	Total Weight Solution (lb/IG)	NaOH (Dry g/liter)	Total Weight Solution (g/liter)	NaOH (Dry lb/ft³)	Total Weight Solution (lb/ft³)
30	23.25	1.328	35.8	65.6	3.33	11.06	3.99	13.28	398.4	1,325	24.87	82.74
32	24.80	1.349	37.5	69.8	3.60	11.24	4.33	13.49	431.7	1,346	26.95	84.05
34	26.35	1.370	39.1	73.9	3.89	11.41	4.67	13.70	463.7	1,367	29.07	85.33
36	27.90	1.390	40.7	78.0	4.18	11.58	5.02	13.90	500.4	1,387	31.24	86.60
38	29.45	1.410	42.2	82.0	4.47	11.74	5.37	14.10	538.8	1,407	33.45	87.86
40	31.00	1.430	43.6	86.0	4.77	11.91	5.73	14.30	572.0	1,427	35.71	89.10
42	32.55	1.449	45.0	89.9	5.08	12.07	6.10	14.50	608.7	1,447	38.00	90.31
44	34.10	1.469	46.3	93.7	5.39	12.23	6.48	14.69	646.1	1,466	40.34	91.50
46	35.65	1.487	47.5	97.5	5.71	12.39	6.86	14.88	684.2	1,485	42.71	92.67
48	37.20	1.507	48.8	101.3	6.04	12.55	7.25	15.07	723.1	1,504	45.14	93.86
50	38.75	1.525	49.9	105.1	6.37	12.70	7.64	15.26	762.7	1,522	47.61	95.03

Source: Speight, James, Lange's Handbook of Chemistry, 16th Ed. Copyright © 2005 McGraw-Hill. Reproduced with permission of The McGraw-Hill Companies.

Table C-7   Specific gravity, boiling points, and freezing points of various aqua ammonia solutions

°Baumé 60°F	Sp. Gr. 60°/60°F	Ammonia Concentration			Approximate Boiling Pt, °F at One Atmos of Pres.	Approximate Freezing Pt,°F
		%NH$_3$ by Wt. in Solution	Lb NH$_3$ per ft^3 Solution	Lb NH$_3$ per Gal. Solution		
10	1.0000	0.00	0.0000	0.0000	212	+32
11	0.9929	1.62	0.7504	0.1003	195	+28
12	0.9859	3.30	2.0289	0.2712	186	+25
13	0.9790	5.02	3.0652	0.4097	177	+20
14	0.9722	6.74	4.0871	0.5463	171	+16
15	0.9655	8.49	5.1122	0.6834	163	+10
16	0.9589	10.28	6.1479	0.8218	156	+6
17	0.9524	12.10	7.1873	0.9608	149	−2
18	0.9459	13.96	8.2355	1.1009	142	−11
19	0.9396	15.84	9.2824	1.2409	134	−15
20	0.9333	17.76	10.337	1.3819	127	−26
21	0.9272	19.68	11.380	1.5213	120	−31
22	0.9211	21.60	12.408	1.6587	111	−46
23	0.9150	23.52	13.422	1.7942	103	−56
24	0.9091	25.48	14.446	1.9312	95	−69
25	0.9032	27.44	15.456	2.0662	88	−89
26	0.8974	29.40	16.454	2.1996	85	−110
27	0.8917	31.36	17.439	2.3313	73	−123
28	0.8861	33.32	18.413	2.4615	66	−148
29	0.8805	35.28	19.375	2.5901	59	−143

Source: Speight, James, *Lange's Handbook of Chemistry*, 16th Ed. Copyright © 2005 McGraw-Hill. Reproduced with permission of The McGraw-Hill Companies.

Appendix D

Calculator Derivations

The following calculators are presented in this field guide. They are listed in this appendix for convenience. The detailed derivation of the formulas is shown as well as the calculator as it appears in the body of the field guide. It should be noted that the calculators in the field guide are stripped of the details and only show a single constant that may be the combination of several. The indicated chapters have detailed instructions and examples of how to correctly use the calculators.

Gas Chemical Feed Calculators

Gas Feeder Setting Calculators (US and SI units)

c4-1 Gas feeder setting (lb/d) = 8.34 lb/MG/mg/L × dosage (mg/L) × water flow (mgd)

$$\boxed{\text{Feeder setting} \atop \text{(lb/d)}} = 8.34 \times \boxed{\text{Dosage} \atop \text{(mg/L)}} \times \boxed{\text{Water flow (mgd)}}$$

c4-2 Gas feeder setting (g/hr) = d/24 hr × g/1,000 mg × dosage (mg/L) × water flow (ML/d is 1,000,000L/d)

$$\boxed{\text{Feeder setting} \atop \text{(g/hr)}} = 41.7 \times \boxed{\text{Dosage} \atop \text{(mg/L)}} \times \boxed{\text{Water flow} \atop \text{(ML/d)}}$$

Ozone Dosage Calculators (Oxygen Feed System)

Table D-1 illustrates a simplified calculation of the ozone production and dosage for a specific situation.

Table D-1   Ozone dosage calculation spreadsheet example: oxygen feed

Constants				
Molar Volume	0.08205	L/mole/deg K		
Absolute Temperature	273.15	°K		
Constant	453.6	g/lb		
Constant	28.32	L/ft^3		
Constant	8.34	lb/gal		
Gram Molecular Weight (GMW) of Gas				
Ozone	48	g/mole		
Oxygen	32	g/mole		
Nitrogen	28	g/mole		
Argon	40	g/mole		
Standard Temperature	20	°C		
Molar Volume	24.05296	L/mole		
Feed-gas Composition			**Feed-gas (GMW)**	
Oxygen	98	%vol	31.36	g/mole
Nitrogen	2	%vol	0.56	g/mole
Argon		%vol		g/mole
		Feed-gas gram molecular weight	31.92	g/mole
Feed-gas Density	1.327072	g/L		
	0.0829	lb/ft^3		
Ozone Operating Conditions				
Gas Flow	1	scfm	1.698	Nm^3/hr
Ozone Concentration	1	%wt		
Water Flow Rate	1	mgd	3.785	ML/d
Ozone Calculations				
Ozone Production	1	lb/d	0.541668	kg/d
Ozone Dosage	0.14	mg/L		

NOTE: This spreadsheet is provided for illustration and to show constants used in the calculators. Formulas embedded in the spreadsheet are not functional in this example.

c4-3   Ozone dosage (mg/L) = air flow (scfm) × gas density (0.829 lb/ft^3) × wt%/10wt% × 1,440 min/d ÷ 100% ÷ water flow (mgd) ÷ g/1,000 mg

$$\boxed{\begin{array}{c}\text{Ozone}\\\text{dosage}\\\text{(mg/L)}\end{array}} = 0.14 \times \boxed{\begin{array}{c}\text{\%wt ozone}\\\text{concentration}\end{array}} \times \boxed{\begin{array}{c}\text{Gas flow}\\\text{(scfm)}\end{array}} \div \boxed{\begin{array}{c}\text{Water flow}\\\text{(mgd)}\end{array}}$$

c4-4   Ozone dosage (mg/L) = air flow (Nm^3/hr) × gas density (1.33 g/L) × wt%/10wt% × 1,440 min/d ÷ 100% ÷ water flow (ML/d)

$$\boxed{\begin{array}{c}\text{Ozone}\\\text{dosage}\\\text{(mg/L)}\end{array}} = 0.32 \times \boxed{\begin{array}{c}\text{\%wt ozone}\\\text{concentration}\end{array}} \times \boxed{\begin{array}{c}\text{Gas flow}\\\text{(Nm}^3\text{/hr)}\end{array}} \div \boxed{\begin{array}{c}\text{Water flow}\\\text{(ML/d)}\end{array}}$$

Ozone Generator Gas Flow Calculators (Oxygen Feed System)

c4-5   Ozone generator gas flow (scfm) = ozone dosage (mg/L) × water flow (mgd) ÷ ozone concentration in gas feed (wt%/10wt%) × 8.34 lb/gal × 454 g/lb × g/1,000 mg

$$\boxed{\begin{array}{c}\text{Gas}\\\text{flow}\\\text{(scfm)}\end{array}} = \boxed{\begin{array}{c}\text{Ozone}\\\text{dosage}\\\text{(mg/L)}\end{array}} \times \boxed{\begin{array}{c}\text{Water flow}\\\text{(mgd)}\end{array}} \div 0.14 \div \boxed{\begin{array}{c}\text{\%wt ozone}\\\text{concentration}\end{array}}$$

c4-6   Ozone generator gas flow (Nm^3/hr) = ozone dosage (mg/L) × water flow (ML/d) ÷ ozone concentration in gas feed (wt%/10wt%) × d/24 hr × Nm^3/1,000 ML

$$\boxed{\begin{array}{c}\text{Gas flow}\\\text{(Nm}^3\text{/hr)}\end{array}} = \boxed{\begin{array}{c}\text{Ozone}\\\text{dosage}\\\text{(mg/L)}\end{array}} \times \boxed{\begin{array}{c}\text{Water}\\\text{flow}\\\text{(ML/d)}\end{array}} \div 0.32 \div \boxed{\begin{array}{c}\text{\%wt ozone}\\\text{concentration}\end{array}}$$

Ozone Dosage Calculators (Air Feed System)

Table D-2 contains dosage results for a set of common conditions and where air is the feed gas to the generator.

Table D-2   Ozone dosage calculation spreadsheet example: air feed

Constants				
Molar Volume	0.08205	L/mole/deg K		
Absolute Temperature	273.15	°K		
Constant	453.6	g/lb		
Constant	28.32	L/ft³		
Constant	8.34	lb/gal		
GMW of Gas				
Ozone	48	g/mole		
Oxygen	32	g/mole		
Nitrogen	28	g/mole		
Argon	40	g/mole		
Standard Temperature	20	°C		
Molar Volume	24.05296	L/mole		
Feed-gas Composition			**Feed-gas GMW**	
Oxygen	20.94	%vol	6.7008	g/mole
Nitrogen	78.12	%vol	21.8736	g/mole
Argon	0.937	%vol	0.3748	g/mole
		Feed-gas GMW	28.9492	g/mole
Feed-gas Density	1.203561	g/L		
	0.0751	lb/ft³		
Ozone Operating Conditions				
Gas Flow	1	scfm	1.698	Nm³/hr
Ozone Concentration	1	%wt		
Water Flow Rate	1	mgd	3.785	ML/d
Ozone Calculations				
Ozone Production	1	lb/d	0.491255	kg/d
Ozone Dosage	0.13	mg/L		

c4-7 Ozone dosage (mg/L) = air flow (scfm) × gas density (0.075 lb/ft^3) × wt%/2wt% × 1,440 min/d ÷ 100% ÷ water flow (mgd) ÷ g/1,000 mg

$$\boxed{\text{Ozone dosage (mg/L)}} = 0.13 \times \boxed{\text{\%wt ozone concentration}} \times \boxed{\text{Gas flow (scfm)}} \div \boxed{\text{Water flow (mgd)}}$$

c4-8 Ozone dosage (mg/L) = air flow (Nm^3/hr) × gas density (1.20 g/L) × wt%/2wt% × 1,440 min/d ÷ 100 % ÷ water flow (ML/d)

$$\boxed{\text{Ozone dosage (mg/L)}} = 0.29 \times \boxed{\text{\%wt ozone concentration}} \times \boxed{\text{Gas flow (Nm}^3\text{/hr)}} \div \boxed{\text{Water flow (ML/d)}}$$

Ozone Generator Gas Flow Calculators (Air Feed System)

c4-9 Ozone generator gas flow (scfm) = ozone dosage (mg/L) × water flow (mgd) ÷ ozone concentration in gas feed (wt%/2wt%) × 8.34 lb/gal × 454 g/lb × g/1,000 mg

$$\boxed{\text{Gas flow (scfm)}} = \boxed{\text{Ozone dosage (mg/L)}} \times \boxed{\text{Water flow (mgd)}} \div 0.13 \div \boxed{\text{\%wt ozone concentration}}$$

c4-10 Ozone generator gas flow (Nm^3/hr) = ozone dosage (mg/L) × water flow (ML/d) ÷ ozone concentration in gas feed (wt%/2wt%) × d/24hr × Nm^3/1,000 ML

$$\boxed{\text{Gas flow (Nm}^3\text{/hr)}} = \boxed{\text{Ozone dosage (mg/L)}} \times \boxed{\text{Water flow (ML/d)}} \div 0.29 \div \boxed{\text{\%wt ozone concentration}}$$

Chlorine Dioxide Feed Calculators

c4-11 Chlorine dioxide feed (lb/d) = dosage (mg/L) × water flow (mgd) × 8.34 lb/gal

$$\boxed{\text{ClO}_2 \text{ feed (lb/d)}} = 8.34 \times \boxed{\text{Dosage (mg/L)}} \times \boxed{\text{Water flow (mgd)}}$$

c4-12 Chlorine dioxide feed (kg/d) = dosage (mg/L) × water flow (ML/d)

$$\boxed{\text{ClO}_2 \text{ feed (kg/d)}} = \boxed{\text{Dosage (mg/L)}} \times \boxed{\text{Water flow (ML/d)}}$$

Chlorine Dioxide Production Relationship Calculators

c4-13 1.34 wt units sodium chlorite + 0.53 wt units chlorine = 1 wt unit chlorine dioxide

Sodium chlorite (1.34 wt units/d)	+	Chlorine (0.53 wt units/d)	=	ClO_2 production (1 wt units/d)

c4-14 Sodium chlorite needed of given concentration (wt units) = 1.34 wt units/1 wt unit of chlorine dioxide × number of wt units of chlorine dioxide produced ÷ % sodium chlorite solution strength/100

Wt units of sodium chlorite for concentration used	=	1.34 wt units/d/1 wt unit of ClO_2 produced by pure sodium chlorite	×	Number of wt units of ClO_2 produced	÷	Decimal concentration of sodium chlorite

c4-15 Chlorine needed (wt units) = 0.53 wt units/1 wt unit of chlorine dioxide × number of wt units of chlorine dioxide produced ÷ % chlorine concentration/100

Wt units of chlorine for concentration used	=	0.53 wt units/d/1 wt unit of ClO_2 produced by pure chlorine	×	Number of wt units of ClO_2 produced	÷	Decimal concentration of chlorine

Chlorine Dioxide Generator Output Concentration Calculators

c4-16 Chlorine dioxide generator output concentration (mg/L) = chlorine dioxide production (lb/d) ÷ water flow through generator (gpm) × 454 g/lb × 1,000 mg/g ÷ 1,440 min/d × gal/3.785 L

Generator ClO_2 concentration (mg/L)	= 82.6 ×	ClO_2 production (lb/d)	÷	Water flow through the generator (gpm)

c4-17 Chlorine dioxide generator output concentration (mg/L) = chlorine dioxide production (kg/d) ÷ water flow through generator (L/m) × 1,000,000 mg/kg ÷ 1,440 min/d

Generator ClO_2 concentration (mg/L)	= 694.4 ×	ClO_2 production (kg/d)	÷	Water flow through the generator (L/min)

c4-18 Chlorine dioxide generator output concentration (mg/L) = dosage (mg/L) × water flow (mgd) ÷ water flow through generator (gpm) × d/1,440 min

$$
\boxed{\begin{array}{c}\text{Generator}\\ \text{ClO}_2\\ \text{concentration}\\ \text{(mg/L)}\end{array}} = 694.4 \times \boxed{\begin{array}{c}\text{Dosage}\\ \text{(mg/L)}\end{array}} \times \boxed{\begin{array}{c}\text{Water}\\ \text{flow}\\ \text{(mgd)}\end{array}} \div \boxed{\begin{array}{c}\text{Water flow}\\ \text{through the}\\ \text{generator}\\ \text{(gpm)}\end{array}}
$$

c4-19 Chlorine dioxide generator output concentration (mg/L) = dosage (mg/L) × water flow (ML/d) ÷ water flow through generator (L/m) × d/1,440 min

$$
\boxed{\begin{array}{c}\text{Generator}\\ \text{ClO}_2\\ \text{concentration}\\ \text{(mg/L)}\end{array}} = 694.4 \times \boxed{\begin{array}{c}\text{Dosage}\\ \text{(mg/L)}\end{array}} \times \boxed{\begin{array}{c}\text{Water}\\ \text{flow}\\ \text{(ML/d)}\end{array}} \div \boxed{\begin{array}{c}\text{Water flow}\\ \text{through the}\\ \text{generator}\\ \text{(L/min)}\end{array}}
$$

Chlorine Dioxide Generator Output Purity

c4-20 Purity of generator output (%) = 100 × measured generator output concentration (mg/L) ÷ theoretical generator output (mg/L)

$$
\text{Purity (\%)} = \boxed{\begin{array}{c}\text{Measured ClO}_2\\ \text{concentration of}\\ \text{the generator}\end{array}} \div \boxed{\begin{array}{c}\text{Theoretical ClO}_2\\ \text{concentration}\\ \text{from c4-17}\end{array}} \times \ 100
$$

Dry Chemical Feed Calculators

Dry Chemical Feeder Calibration Calculators

c5-1 Feed rate for calibration (lb/hr) = 60 min/hr × weight of chemical (lb) ÷ collection time (min)

$$
\boxed{\begin{array}{c}\text{Feed rate}\\ \text{(lb/hr)}\end{array}} = 60 \times \boxed{\begin{array}{c}\text{Weight of}\\ \text{chemical}\\ \text{(lb)}\end{array}} \div \boxed{\begin{array}{c}\text{Collection time}\\ \text{(min)}\end{array}}
$$

c5-2 Feed rate for calibration (kg/hr) = 60 min/hr × weight of chemical (kg) ÷ collection time (min)

$$
\boxed{\begin{array}{c}\text{Feed rate}\\ \text{(kg/hr)}\end{array}} = 60 \times \boxed{\begin{array}{c}\text{Weight of}\\ \text{chemical}\\ \text{(kg)}\end{array}} \div \boxed{\begin{array}{c}\text{Collection time}\\ \text{(min)}\end{array}}
$$

Dry Chemical Feed Rate Calculators

c6-1   Feed rate (lb/d) = 8.34 lb/MG/(mg/L) × dosage (mg/L) × water flow (MG/d)

$$\boxed{\begin{array}{c}\text{Feed Rate}\\\text{(lb/d)}\end{array}} = 8.34 \times \boxed{\text{Dosage (mg/L)}} \times \boxed{\begin{array}{c}\text{Water flow}\\\text{(mgd)}\end{array}}$$

c6-2   Feed rate (g/hr) = 1 g/1,000 mg × 1d/24 hr × 1,000,000L/ML × dosage (mg/L) × water flow (ML/d)

$$\boxed{\begin{array}{c}\text{Feed rate}\\\text{(g/hr)}\end{array}} = 41.7 \times \boxed{\text{Dosage (mg/L)}} \times \boxed{\begin{array}{c}\text{Water flow}\\\text{(ML/d)}\end{array}}$$

c6-3 $$\boxed{\begin{array}{c}\text{Feed rate}\\\text{(lb/day)}\end{array}} = 834 \times \boxed{\begin{array}{c}\text{Dosage}\\\text{(mg/L)}\end{array}} \times \boxed{\begin{array}{c}\text{Water flow}\\\text{(mgd)}\end{array}} \div \boxed{\begin{array}{c}\text{\% available}\\\text{active}\\\text{ingredient}\end{array}}$$

c6-4 $$\boxed{\begin{array}{c}\text{Feed rate}\\\text{(g/hr)}\end{array}} = 4{,}170 \times \boxed{\begin{array}{c}\text{Dosage}\\\text{(mg/L)}\end{array}} \times \boxed{\begin{array}{c}\text{Water flow}\\\text{(ML/day)}\end{array}} \div \boxed{\begin{array}{c}\text{\% available}\\\text{active}\\\text{ingredient}\end{array}}$$

Dry Chemical Needed To Prepare % Solutions

c6-5   Amount of dry chemical (lb) = 8.34 lb/gal × 1/100% × volume (gal) × % solution strength

$$\boxed{\begin{array}{c}\text{lb of dry}\\\text{chemical}\\\text{needed}\end{array}} = 0.0834 \times \boxed{\begin{array}{c}\text{Number of}\\\text{gallons to be}\\\text{prepared}\end{array}} \times \boxed{\begin{array}{c}\text{\% solution strength}\\\text{needed}\end{array}}$$

c6-6   Amount of dry chemical (kg) = 1/100% × volume (L) × % solution strength

$$\boxed{\begin{array}{c}\text{kg of dry}\\\text{chemical}\\\text{needed}\end{array}} = 0.01 \times \boxed{\begin{array}{c}\text{Number of}\\\text{liters to be}\\\text{prepared}\end{array}} \times \boxed{\begin{array}{c}\text{\% solution strength}\\\text{needed}\end{array}}$$

Liquid Chemical Feed Calculators

Liquid Chemical Pump Rate Calibration Calculator

c7-1   Pump rate (mL/min) = 60 sec/min × volume (mL) ÷ time (sec)

$$60 \times \boxed{\text{Volume pumped (mL)}} \div \boxed{\text{Time (sec)}} = \boxed{\text{Pump rate (mL/min)}}$$

Liquid Chemical Feed Rate Calculators

c8-1   Feed rate (lb/d) = 8.34 lb/mil gal/(mg/L) × dosage (mg/L) × water flow (mgd)

$$\boxed{\begin{array}{c}\text{Feed rate}\\(\text{lb/d})\end{array}} = 8.34 \times \boxed{\begin{array}{c}\text{Dosage}\\(\text{mg/L})\end{array}} \times \boxed{\begin{array}{c}\text{Water flow}\\(\text{mgd})\end{array}}$$

c8-2   Feed rate (g/hr) = 1 g/1,000 mg × 1 d/24 hr × 1,000,000 L/ML × dosage (mg/L) × water flow (ML/d)

$$\boxed{\begin{array}{c}\text{Feed rate}\\(\text{g/hr})\end{array}} = 41.7 \times \boxed{\begin{array}{c}\text{Dosage}\\(\text{mg/L})\end{array}} \times \boxed{\begin{array}{c}\text{Water flow}\\(\text{ML/d})\end{array}}$$

c8-3   Feed rate (gal/d) = 1/8.34 lb/gal × feed rate (lb/d) ÷ specific gravity

$$\boxed{\begin{array}{c}\text{Feed rate}\\(\text{gpd})\end{array}} = 0.12 \times \boxed{\begin{array}{c}\text{Feed rate}\\(\text{lb/d})\end{array}} \div \boxed{\begin{array}{c}\text{Specific}\\\text{gravity}\end{array}}$$

c8-4   Feed rate (L/d) = 1/1,000 g/L × 24 hr/d × feed rate (g/hr) ÷ specific gravity

$$\boxed{\begin{array}{c}\text{Feed rate}\\(\text{L/d})\end{array}} = 0.024 \times \boxed{\begin{array}{c}\text{Feed rate}\\(\text{g/hr})\end{array}} \div \boxed{\begin{array}{c}\text{Specific}\\\text{gravity}\end{array}}$$

Liquid Chemical Specific Gravity Measurement Calculator

c8-5   [chemical weight of 100 mL (g)] ÷ 100 = specific gravity (estimate)

Calculator instructions: Weigh a graduated cylinder, add 100 mL of chemical (exactly), weigh again, subtract weight of cylinder from total; this is the weight of 100 mL of chemical. The weight of 100 mL of water is 100 g. The specific gravity is the weight of the chemical divided by the weight of water.

Solution Preparation Calculators

c8-6   Dry chemical weight (lb) = 1/100% × 8.34 lb/gal × solution strength (%) × volume (gal)

$$\boxed{\begin{array}{c}\text{Dry chemical}\\(\text{lb})\end{array}} = 0.0834 \times \boxed{\begin{array}{c}\text{Solution}\\\text{strength (\%)}\end{array}} \times \boxed{\text{Volume (gal)}}$$

c8-7   Dry chemical weight (kg) = 1/100% × solution strength (%) × volume (L)

$$\boxed{\begin{array}{c}\text{Dry chemical}\\(\text{kg})\end{array}} = 0.01 \times \boxed{\begin{array}{c}\text{Solution}\\\text{strength (\%)}\end{array}} \times \boxed{\text{Volume (L)}}$$

c8-8 Volume (gal) = 8.34 lb/gal × 1/100% × solution strength (%) × volume (gal) ÷ density (lb/gal)

Liquid chemical volume (gal)		Solution strength (%)		Volume (gal)		Density of chemical (lb/gal)
	= 0.0834 ×		×		÷	

c8-9 Volume (L) = 1,000 mL/L × 1/100% × 1 g/ml × solution strength (%) × volume (L) ÷ density (g/L)

Liquid chemical volume (L)		Solution strength (%)		Volume (L)		Density of chemical (g/L)
	= 10 ×		×		÷	

Liquid Chemical Feed Rate Calculators

c8-10 Chemical feed rate (mL/min) = 3.785 L/gal × 1,000 mL/L ÷ 1,440 min/d × 8.34 lb/gal × 100%/1 × chemical dose (mg/L) × water flow (mgd) ÷ solution strength (%) ÷ density of chemical (lb/gal)

Chemical feed rate (mL/min)		Chemical dosage (mg/L)		Water flow (mgd)		Solution strength (%)		Density of chemical (lb/gal)
	= 2,192 ×		×		÷		÷	

c8-11 Chemical feed rate (mL/min) = 1 d/ 1,440 min × 1,000 mL/L × 1 mL/g × 100%/1 × chemical dose (mg/L) × water flow (ML/d) ÷ solution strength (%) ÷ density of chemical (g/mL)

Chemical feed rate (mL/min)		Chemical dosage (mg/L)		Water flow (ML/d)		Solution strength (%)		Density of chemical (g/ML)
	= 69.4 ×		×		÷		÷	

c8-12 Chemical feed rate (mL/min) = 3.785 L/gal × 1,000 mL/L ÷ 1,440 min/d × 100%/1 × chemical dose (mg/L) × water flow (mgd) ÷ solution strength (%) ÷ specific gravity of solution

Chemical feed rate (mL/min)		Chemical dosage (mg/L)		Water flow (mgd)		Solution strength (%)		Specific gravity of chemical solution
	= 262.9 ×		×		÷		÷	

c8-13 Chemical feed rate (mL/min) = 1 d/1,440 min × 100%/1 × 1,000 mL/g chemical dose (mg/L) × water flow (ML/d) ÷ solution strength (%) ÷ specific gravity of solution

Chemical feed rate (mL/min)		Chemical dosage (mg/L)		Water flow (ML/d)		Solution strength (%)		Specific gravity of chemical solution
	= 69.4 ×		×		÷		÷	

Feed Rate Calculators for Very Dilute Solutions

c8-14 Chemical feed rate (mL/min) = 3.785 L/gal × 1,000 mL/L ÷ 1,440 min/d × 100%/1 × chemical dose (mg/L) × water flow (mgd) ÷ solution strength (%)
    Assume specific gravity is that for water

Chemical feed rate (mL/min)		Chemical dosage (mg/L)		Water flow (mgd)		Solution strength (%)
	= 262.9 ×		×		÷	

c8-15 Chemical feed rate (mL/min) = 1 d/1,440 min × 100%/1 × 1,000 mL/g chemical dose (mg/L) × water flow (ML/d) ÷ solution strength (%) ÷ specific gravity of solution
    Assume specific gravity is that for water

Chemical feed rate (mL/min)		Chemical dosage (mg/L)		Water flow (ML/d)		Solution strength (%)
	= 69.4 ×		×		÷	

c8-16 Chemical feed rate (mL/min) = 3.785 L/gal × 1,000 mL/L × 10,000 mg/L/1% ÷ 1,440 min/d × 100%/1 × chemical dose (mg/L) × water flow (mgd) ÷ solution strength (mg/L)
    Assume specific gravity is that for water

Chemical feed rate (mL/min)		Chemical dosage (mg/L)		Water flow (mgd)		Solution strength (mg/L)
	= 2,629,000 ×		×		÷	

c8-17 Chemical feed rate (mL/min) = 1 d/1,440 min × 100%/1 × 1,000 mL/g chemical dose (mg/L) × water flow (ML/d) × 10,000 mg/L/1% ÷ solution strength (mg/L)
    Assume specific gravity is that for water

Chemical feed rate (mL/min)		Chemical dosage (mg/L)		Water flow (ML/d)		Solution strength (mg/L)
	= 694,000 ×		×		÷	

Liquid Chemical Dilution Calculators

c8-18 % strength of diluted solution = % strength of bulk chemical × (volume bulk chemical/total volume of diluted solution)

% strength of diluted solution	=	% strength of bulk chemical solution	×	Dilution ratio (volume bulk chemical/total volume of diluted solution)

c8-19 Bulk chemical used to make diluted solution

Amount of bulk chemical from bulk tank = amount of bulk chemical in dilute tank so:

% strength of bulk solution × volume or feed rate of bulk chemical = % strength of dilute solution × volume or feed rate of diluted solution

c8-19	% strength of diluted solution	×	Volume or feed rate of diluted solution	=	% strength of bulk chemical solution	×	Volume or feed rate of chemical solution

Appendix E

Atomic Weights of Elements

List of Elements

Name	Symbol	Atomic Number	Atomic Weight
Actinium	Ac	89	[227]
Aluminum	Al	13	26.981538
Americium	Am	95	[243]
Antimony	Sb	51	121.76
Argon	Ar	18	39.948
Arsenic	As	33	74.9216
Astatine	At	85	[210]
Barium	Ba	56	137.327
Berkelium	Bk	97	[247]
Beryllium	Be	4	9.012182
Bismuth	Bi	83	208.98038
Bohrium	Bh	107	[264]
Boron	B	5	10.811
Bromine	Br	35	79.904
Cadmium	Cd	48	112.411
Caesium	Cs	55	132.90545
Calcium	Ca	20	40.078
Californium	Cf	98	[251]
Carbon	C	6	12.0107
Cerium	Ce	58	140.116
Chlorine	Cl	17	35.4527
Chromium	Cr	24	51.9961
Cobalt	Co	27	58.9332
Copper	Cu	29	63.546
Curium	Cm	96	[247]
Dubnium	Db	105	[262]

Table continued on next page

List of Elements

Name	Symbol	Atomic Number	Atomic Weight
Dysprosium	Dy	66	162.5
Einsteinium	Es	99	[252]
Erbium	Er	68	167.26
Europium	Eu	63	151.964
Fermium	Fm	100	[257]
Fluorine	F	9	18.9984032
Francium	Fr	87	[223]
Gadolinium	Gd	64	157.25
Gallium	Ga	31	69.723
Germanium	Ge	32	72.61
Gold	Au	79	196.96655
Hafnium	Hf	72	178.49
Hassium	Hs	108	[269]
Helium	He	2	4.002602
Holmium	Ho	67	164.93032
Hydrogen	H	1	1.00794
Indium	In	49	114.818
Iodine	I	53	126.90447
Iridium	Ir	77	192.217
Iron	Fe	26	55.845
Krypton	Kr	36	83.8
Lanthanum	La	57	138.9055
Lawrencium	Lr	103	[262]
Lead	Pb	82	207.2
Lithium	Li	3	[6.941]
Lutetium	Lu	71	174.967
Magnesium	Mg	12	24.305
Manganese	Mn	25	54.938049
Meitnerium	Mt	109	[268]
Mendelevium	Md	101	[258]
Mercury	Hg	80	200.59
Molybdenum	Mo	42	95.94

Table continued on next page

List of Elements

Name	Symbol	Atomic Number	Atomic Weight
Neodymium	Nd	60	144.24
Neon	Ne	10	20.1797
Neptunium	Np	93	[237]
Nickel	Ni	28	58.6934
Niobium	Nb	41	92.90638
Nitrogen	N	7	14.00674
Nobelium	No	102	[259]
Osmium	Os	76	190.23
Oxygen	O	8	15.9994
Palladium	Pd	46	106.42
Phosphorus	P	15	30.973762
Platinum	Pt	78	195.078
Plutonium	Pu	94	[244]
Polonium	Po	84	[210]
Potassium	K	19	39.0983
Praseodymium	Pr	59	140.90765
Promethium	Pm	61	[145]
Protactinium	Pa	91	231.03588
Radium	Ra	88	[226]
Radon	Rn	86	[222]
Rhenium	Re	75	186.207
Rhodium	Rh	45	102.9055
Rubidium	Rb	37	85.4678
Ruthenium	Ru	44	101.07
Rutherfordium	Rf	104	[261]
Samarium	Sm	62	150.36
Scandium	Sc	21	44.95591
Seaborgium	Sg	106	[266]
Selenium	Se	34	78.96
Silicon	Si	14	28.0855
Silver	Ag	47	107.8682
Sodium	Na	11	22.98977

Table continued on next page

List of Elements

Name	Symbol	Atomic Number	Atomic Weight
Strontium	Sr	38	87.62
Sulphur	S	16	32.066
Tantalum	Ta	73	180.9479
Technetium	Tc	43	[98]
Tellurium	Te	52	127.6
Terbium	Tb	65	158.92534
Thallium	Tl	81	204.3833
Thorium	Th	90	232.0381
Thulium	Tm	69	168.93421
Tin	Sn	50	118.71
Titanium	Ti	22	47.867
Tungsten	W	74	183.84
Ununbium	Uub	112	[277]
Ununnillium			
Ununhexium			
Ununoclium			
Ununquadium			
Unununium			
Darmstadtium	Ds	110	[269]
Roentgenium	Rg	111	[272]
Uranium	U	92	238.0289
Vanadium	V	23	50.9415
Xenon	Xe	54	131.29
Ytterbium	Yb	70	173.04
Yttrium	Y	39	88.90585
Zinc	Zn	30	65.39
Zirconium	Zr	40	91.224

References

Baruth, Edward E., ed. 2005. *Water Treatment Plant Design*. 4th ed. American Water Works Association and American Society of Civil Engineers. New York: McGraw-Hill, Inc.

Casale, Robin J. 2001. Improving Chemical Handling Procedures Can Help Reduce Associated Treatment Problems. *Journal AWWA* 93, Part 9 (September): 95–106.

Chlorine Institute, The. 1999. *Water and Wastewater Operators' Chlorine Handbook*. Washington, D.C.: The Chlorine Institute.

Crittenden, John C., Montgomery Watson Harza, et al. 2005. *Water Treatment Principles and Design*. 2nd ed. Hoboken, N.J.: John Wiley & Sons Inc.

Eaton, Andrew D., and M.A.H. Franson, eds. 2005. *Standard Methods for the Examination of Water and Wastewater*. 21st ed. Washington, D.C.: American Public Health Association.

Gates, Donald. 1998. *The Chlorine Dioxide Handbook*. Denver, Colo.: American Water Works Association.

Giorgi, John. 2007. *Math for Water Treatment Operators*. Denver, Colo.: American Water Works Association.

HDR Engineering Inc. *Handbook of Public Water Systems*. 2001. 2nd ed. New York: John Wiley & Sons Inc.

Kawamura, Susumu. 2000. *Integrated Design and Operation of Water Treatment Facilities*. 2nd ed. New York: John Wiley & Sons Inc.

Kerri, Kenneth D.; California State University, Sacramento, Office of Water Programs; National Environmental Training Association; California Sanitary Engineering Branch; and United States Environmental Protection Agency Office of Drinking Water. 2004. *Water Treatment Plant Operation: A Field Study Training Program*. Vols. 1 (5th ed.) and 2 (4th ed.). Sacramento, Calif.: California State University.

Lauer, William C., Timothy J. McCandless, and John M. Stubbart. 2004. *AWWA Water Operator Field Guide*. Denver, Colo.: American Water Works Association.

Letterman, Raymond D., ed. 1999. *Water Quality and Treatment*. 5th ed. New York: McGraw-Hill Inc.

MacPhee, Michael, David A. Cornwell, and Richard Brown. 2002. *Trace Contaminants in Drinking Water Chemicals*, Denver, Colo.: Awwa Research Foundation.

Pizzi, Nicholas G. 2005. *Water Treatment Operator Handbook*. Rev. ed. Denver, Colo.: American Water Works Association.

Rakness, Kerwin L. 2005. *Ozone in Drinking Water Treatment: Process Design, Operation, and Optimization.* Denver, Colo.: American Water Works Association.

Speight, J. G., and Norbert Adolph Lange. 2004. *Lange's Handbook of Chemistry.* 16th ed. Maidenhead, U.K.: McGraw-Hill Professional.

United States Environmental Protection Agency. 1999. Chlorine Dioxide. Chapter 4 of *Alternative Disinfectants and Oxidants Guidance Manual.* Washington, D.C.: USEPA.

Water Treatment. 2003. 3rd ed. Denver, Colo.: American Water Works Association.

White, George Clifford. 1999. *Handbook of Chlorination and Alternative Disinfectants.* 4th ed. New York: John Wiley & Sons.

Index

Note: *f.* indicates figure; *t.* indicates table.

www.ingramcontent.com/pod-product-compliance
Lightning Source LLC
Chambersburg PA
CBHW070725220326
41598CB00024BA/3303